The Philosophy and Medicine series is dedicated to publishing monographs and collections of essays that contribute importantly to scholarship in bioethics and the philosophy of medicine. The series addresses the full scope of issues in bioethics and philosophy of medicine, from euthanasia to justice and solidarity in health care, and from the concept of disease to the phenomenology of illness. The Philosophy and Medicine series places the scholarship of bioethics within studies of basic problems in the epistemology, ethics, and metaphysics of medicine. The series seeks to publish the best of philosophical work from around the world and from all philosophical traditions directed to health care and the biomedical sciences. Since its appearance in 1975, the series has created an intellectual and scholarly focal point that frames the field of the philosophy of medicine and bioethics. From its inception, the series has recognized the breadth of philosophical concerns made salient by the biomedical sciences and the health care professions. With over one hundred and twenty five volumes in print, no other series offers as substantial and significant a resource for philosophical scholarship regarding issues raised by medicine and the biomedical sciences.

More information about this series at http://www.springer.com/series/6414

Philosophy and Medicine

Volume 138

Jean-Pierre Cléro

Reflections on Medical Ethics

A Search for Categories of Medical Ethics

Springer

Jean-Pierre Cléro
Department of Philosophy
University of Rouen
Paris, France

ISSN 0376-7418 ISSN 2215-0080 (electronic)
Philosophy and Medicine
ISBN 978-3-030-65232-6 ISBN 978-3-030-65233-3 (eBook)
https://doi.org/10.1007/978-3-030-65233-3

This Springer imprint is published by the registered company Springer Nature Switzerland AG
The registered company address is: Gewerbestrasse 11, 6330 Cham, Switzerland

Preface

This book is the follow-up to research begun in French in 2011 and entitled *Calcul Moral ou Comment Raisonner en Ethique* (Moral Calculus – How to Reason in Ethics), and followed, in English, in 2018, under the title *Rethinking Medical Ethics*, published by Ibidem (Stuttgart). This research consists, *firstly*, in the criticism of the traditional categories of ethics that medical circles and medical ethics rarely question, even though they introduce contradictions in problems they should help to solve. This is especially the case of Kantian concepts, those of *person*, *personality*, *consent* and *dignity* which, rather than offering solutions to conflicts that entail ethical questions, seem to be transformed and bent by them. It consists, *secondly*, in looking for categories that would be more efficient, ethically speaking. That is what we have done in our more recent research, when we tried to promote *intimacy* which is, quite wrongly, put under the category of the *person* and its *interiority*. We have shown the paradox that, far from being equivalent to the person or being in line with 'personalistic' conceptions of ethics, intimacy would be better thought in a utilitarian version of ethics. This time, it is to the concept of *meeting* that we would like to give an ethical meaning. This has sometimes been ventured, but those who supported it seem to have abandoned the endeavour.

Of course, we are not trying to innovate, for that would be a waste of time in ethics. We only want to avoid the sophisms and paralogisms to which the use of the categories most resorted to today gives rise. When Descartes says that it is in the knowledge of passions, which is so crucial in ethics, that the reason of the ancients is the most defective – more so than in the knowledge of any other domain [2, p. 327] – we cannot but agree with him. We believe so little in a progress of ethical knowledge that we will go back, beyond Kantian ethics, to the category of *meeting* which is quite close to the passion that Descartes called *admiration*, [3, p. 373] and which he thought was the first of all passions in that all the others owe it something. We could not do better in ethics than follow an argumentative movement which sets aside ideas that used to be thought decisive, and which brings back to the foreground ideas that used to be thought to have been superseded by those that we will now set aside. However, ideas that have been set aside are not always so for ever, any more than those that are made to come back were ever completely lost: it is not the identical components that are brought into play in the categories that are being used, although they are given the same names. That is why, although we launch an attack on the concept of *person* and the concepts that necessarily revolve around it, we could not advocate that their paradigm be purely and simply abandoned. Ethics does not work like those sciences which seem to change paradigms and never come back to them. Reversals are always possible, at least we cannot set them aside, if only as trials, in situations in which an ethical position is required, even though the sciences may refine the concepts that ethics uses, as is the case, for example, in probabilities. Admittedly, the exquisite discussion on the fate of *elpis* <ἐλπίς, hope, expectation> in Plato, that of *tuché* <τύχη, meeting> in Aristotle and that of the *paradox* in Diodorus Chronos, are nowadays given, in the works on probabilities that we have been benefitting from since the seventeenth century, an in-depth analysis and flexibility of calculation that they did not previously have. Some categories are being completely reassessed or weighed very differently, but we could not say that those ancient discussions no longer have any meaning today, or that they have meaning only for some historians of ideas.

How could we, when moral, amoral and immoral, religious and non-religious positions – provided they do not contravene the rules of justice – are the very matter of what is being built under the name of *ethics* in a given situation? We are going to show that ethics belongs to the order of diplomacy, that it is neither the same as morals – with what morals could it be mixed to achieve its ends? – nor as religions or non-religions, for the same reason. Though morals may have a strong rational component, as a reading of Kant's works could show, the rationality of diplomacy, which settles in the contingency of situations, cannot be the same. Its rationality is that of calculation, based on the sole requirement that it creates as little unhappiness as possible in situations where one knows that sometimes they do not end well. One sometimes has to accept solutions to situations which are morally outrageous in the same way that they upset our morals: morals, though it feeds ethical debate, like religion, shares only outward affinities with ethics. The latter may even go against some aspects of the law in order for war not to prevail over civil peace. Then it puts two dangers in the scales and allows that which seems less dangerous to carry the day.

It is the link to situations and circumstances which are always singular and which cannot but be compared – without that comparison meaning necessarily that they can be solved identically, but rather similarly – that results in medical ethics being neither the application of some general ethics to singular facts that contain medical elements – one sick individual (or individuals), member(s) of their family (or families), some more or less curable disease, etc. – nor a radically different sector that would have nothing in common with business ethics, or sports ethics, or legal ethics, etc., as if it were possible to juxtapose those ethics without them sharing any commensurability. In reality, any ethical consideration is a circumstantial consideration. The categories on the basis of which ethical difficulties are solved, when they are – and in one sense, it is only when there are difficulties that ethical questions are set – are the thoughts of a particular situation and cannot be used again for another one, even if it is similar, without great precautions. Everything is always to be done all over again in ethics, even though ethics is not without memory.

Even more, and the preceding remarks point in this direction, despite all the efforts we have seen to make ethics become rational, that is, always capable of presenting its reasons – which is, incidentally, what makes it different from morals, which may have good reasons to be sentimentalist, as reason shows that it cannot solve the problems presented to it on its own – ethics is a place to make choices which one knows do not belong only to reason, even though reason would be tempted to push one in the opposite directions. In ethical questions, there are attachments and preferences that are as important as reasons, in particular when the expression of those preferences and attachments does not go against the interests of other individuals or groups. However, the assessment of the preferences, attachments and interests of others, like that of our own, is not without risk. It is reason itself which recommends giving them free rein, provided it calculates them or their probabilities. The calculations of reason do not remove passionate considerations. They only limit them, sometimes challenging them, but also justifying them. Despite all its efforts to argue, ethics does not remove the role of irrational elements but regulates it – not as if it were a last concession it made them, but as the necessary driving force of its own existence. So that ethics seems to be more a rationalization of elements that are not in themselves rational than to be a directly rational construction in its own nature. We will have the possibility, when analysing our ethical behaviours in relation to blood transfusion, for instance, to confirm non-rational intuitions. It may be that one could not mount rational arguments for a position, but that, however, one did give it more value than it seems to be entitled to. One is wary of reason and calculation in ethics. There is a sort of misology of the most argued ethics, and of an ethics that breaks that sort of law by being a rationalization rather than the proper establishment of reason which seems to be cynical, dangerous, or simply inapplicable. One only has to think of the often quite rational positions of P. Singer: they can easily be seen as provocations in circles that are not religious at all. One should think of some of one's attitudes towards the death penalty that reason does not necessarily invalidate: after all, Kant was not necessarily wrong against Beccaria, and those who are against the death penalty often have difficulties in showing why everyone should be against it. Most of those who do not want a 'justice that kills' start from an intuition and develop arguments based on it, but reason is not at their core. It seems that the maximum of rationality that may be required from the exercise of reason in ethics is often to show that it is not possible to solve the problem rationally, as sometimes happens in mathematical problems. One can show that a problem does not have any rational solution, and it is reason which shows it. It seems that the Humean argument in Book III of *A Treatise of Human Nature* is truly the following: reason cannot find its disqualification in facts, but it can disqualify itself, by its own operation, by the very intelligence of what it can do in ethics, as a candidate for the sole foundation of ethical behaviours. Reason can defeat the simple fact of aporia by showing why it cannot overcome the aporia when solving some problems.

This preface would not be complete without our thanks and a double warning to the reader. My thanks are primarily to Miguel Olmos and Benoît Roux, who are, respectively, director and research engineer at the ERIAC research group, to which I am affiliated at Rouen University (France), and who did what could be done for this book to be published by Springer in English. I would also like to thank Isabelle Mulot, who, on behalf of IRISH, has made it possible to pay my translator. I thank Solène Semichon, who has been translating my work into English for a long time and has once more proved herself more than capable of meeting the challenge. Almost all the persons previously mentioned have already played a major role in the translation and publication in English of *Rethinking Medical Ethics*. Thanks to them, my work is given a direction which unifies it and the opportunity to reach an audience it would never reach without their devoted help. Working with ERIAC and IRISH, I have accumulated a series of debts. Indeed, for more than 15 years, Rouen University has made it possible for me to teach numerous courses and has created links with all sorts of foreign universities – Université Laval (in Quebec, with my friend Thierry Belleguic), University of Bucarest (in Romania, with my friends late Valentin Muresan and Dean Romulus Brâncoveanu), University of Mons (in Belgium with my friend Anne Staquet) and University of Brussels (with my atheist friends) in the field of medical ethics. The weaving of ideas would not have been the same without my contact with the Bentham Project of London and especially my friend Michael Quinn's spontaneous proofreading of *Rethinking Medical Ethics*. Without the friends I met in Rouen, Annie Hourcade, Pierre Czernichow and Benoît Misset; without the numerous friends of the Rouen and Rouvray University hospital, with whom I have been working for a long time; and all of whose names I cannot mention, this book would never have been written. I could not forget the invitations to the prestigious symposiums and seminars of Professors Christian Hervé and Emmanuel Hirsch, and to write in the journals they direct, *Ethics, Medicine & Public Health* and *Revue Française d'Éthique Appliquée*, which have allowed me to work in directions in which I would not necessarily have worked myself. Lastly, how could I not thank my long-time friends in the Centre Bentham at Sciences Po-Paris – Guillaume Tusseau, Anne Brunon-Ernst, Emmanuelle de Champs, Claire Wrobel, Malik Bozzo-Rey and Benjamin Bourcier – who have infallibly helped my projects thanks to their intellectual encouraging and material help? I hope the friends I have forgotten will not bear me a grudge! An author always realizes, when he is writing or signing a book, that even though he may be working sincerely and persuaded of its originality, he is but the spokesperson of several people and constantly echoes multiple debates.

I would particularly like to thank my friend Michael Quinn who, invited by Chris Wilby to comment on the text, raised some queries with me and entered into a stimulating dialogue which has considerably ameliorated the expression of the philosophical arguments, and provoked new thoughts in some areas.

Lastly, I need to doubly alert the reader. First, when this book, *New Reflections on Medical Ethics*, is referred to as a follow-up to *Rethinking Medical Ethics*, it does not mean that the former cannot be read independently and without reading the latter. Each time a result of *Rethinking Medical Ethics* is necessary for the understanding of

New Reflections on Medical Ethics, it is indicated and summarized so that there is absolutely no need to refer to the first volume.

Secondly, and much more importantly, though the book essentially focuses on medical ethics, it also mentions other aspects of ethics. Moreover, the expressions *ethics of care*, *medical ethics* and even *bioethics* are not strictly distinguished. That is not because I fail to recognize that there are such divisions as *business ethics*, *ethics of jurists*, *environmental ethics* and other numerous types of ethics. However, those distinctions seem to be nominal ones, having no essential significance for my purpose. That does not mean either that I think there is some general ethics that the philosopher should think about in its abstract nature and that should then be 'applied' – as is sometimes so awkwardly said – to different domains, which would make it possible to distinguish different ethics. Indeed, the reader who has read my first book or will read the second will understand that I think that what Bachelard nicely said of the scientific method could apply to ethics:[1] that it is first and foremost a circumstantial discourse and a behaviour based on which, when trying to structure them and act upon them in their singularity, one finds subjects, actors, agents, patients, objects, aims and motives – the constituents of the conflict that puts them all into action. Undoubtedly, it is not impossible that structuring and solving one conflict may be used to solve others, but it should not be an 'application' of what has been established in one case to another one. Ethics is a sort of casuistry, not a set of rules that should be used to solve specific problems. The judgement on particular conditions is each time at the core of things, and it is never said that the rules produced in one situation could be used in the same way in similar cases, or in cases that seem to be similar. One thus allows oneself to 'go out of' a field, because no field of ethics can have predefined and predetermined limits.

Department of Philosophy
University of Rouen
Paris, France
18th June 2019

Jean-Pierre Cléro Paris

Bibliography

[1] Bachelard, G. (1991). *Le nouvel esprit scientifique*. Paris: Quadrige/PUF.
[2] Descartes, R. (1986). *Les passions de l'âme*, Ière Partie, Art. 1. In *Oeuvres de Descartes*, Adam & Tannery, XI, Vrin, Paris.
[3] Descartes, R. (1986). *Les passions de l'âme*, IIde Partie, Art. LII1. In *Oeuvres de Descartes*, Adam & Tannery, XI, Vrin, Paris.

[1]Bachelard's sentence is: 'Un discours sur la méthode scientifique sera toujours un discours de circonstance, il ne décrira pas une constitution définitive de l'esprit scientifique' [1, p. 139].

Acknowledgements

The present book could not have been written without the combined efforts of Solène Semichon and Michael Quinn. I would like to thank them for that.

Contents

Chapter 1
Of Ethics as Diplomacy

Abstract We are going to show that ethics belong to the register of *diplomacy*, that it is neither the same as moral points of view – with what morals could it be mixed to achieve its ends? – nor as religious or irreligious points of view, for the same reason. Though morals may have a strong rational component, as a reading of Kant's works could show, the rationality of *diplomacy*, which settles in the contingency of situations, cannot be the same. Its rationality is that of calculation, based on the sole requirement that it creates as little unhappiness as possible in situations where one knows that sometimes they do not end well. One sometimes must accept solutions to situations which are morally outrageous in the same way that they upset our morals: morals, though it feeds ethical debate, like religion, share only outward affinities with ethics. The latter may even go against some aspects of the law in order for war not to prevail over civil peace. Then it puts two dangers in the scales and allows that which seems less dangerous to carry the day.

In reality, any ethical consideration is a circumstantial consideration. The categories on the basis of which ethical difficulties are solved, when they are – and in one sense, it is only when there are difficulties that ethical questions are set –, are the thoughts of a particular, singular, contingent situation and cannot be used again for another one, even if it is similar, without great precautions. Nevertheless, we have no reasons to give up the search of rules, which are only less general than moral laws or less imperative than juridical laws.

Keywords Authority of states · Democracy · Diplomacy · Lobbies · Relation between ethics and politics

It is common knowledge that democracies have been in a state of crisis for a few decades and that that state has worsened at a faster pace over the past few years. So-called 'populist' votes which benefit the far right are a symptom of it. Should one lament this phenomenon as if one lived in some idyllic past, when citizens enjoyed freedom, equality and brotherhood? I do not think so; and even though freedom characterizes democracies quite well – citizens in democracies are in general almost as free as they could wish, and are free so long as they remain in a democracy – one

J.-P. Cléro, *Reflections on Medical Ethics*, Philosophy and Medicine 138,
https://doi.org/10.1007/978-3-030-65233-3_1

also knows that democracies have never produced the equality and brotherhood they promised. That may be why those who have never known anything but the bad side of inequality, and the exclusion of any brotherhood, make the citizens of the states where they live pay the price of their despair by voting in a way that threatens even the freedoms they do not care about anymore. This is a very dangerous phenomenon and should not be neglected, but have democracies fallen, or are they falling, into an abyss in the same way as in the 1930s? Certainly not. Something quite different is happening, of which people's current passion for ethics is a decisive sign. We would like to start first by questioning that sign, which is more encouraging than it is discouraging, by linking it to an accelerated weakening of States, perhaps to a crumbling of laws and politics, which makes us lose a good part of our former landmarks. We will set aside the nationalist spurts that we, like everybody, can see, but which do not arouse any hope in anybody, not even in those who claim they are representing them and who know quite well that there is not even the shadow of reality in what they recommend. Worrying though they may be, those spurts do not weigh on the dynamic that is developing under our eyes.

During the 2017 presidential elections, French people quite legitimately made fun of the poor campaign, or at least the unrealistic one, of a party the leaders of which considered a double currency system – a currency for each pocket, as was ironically said. But people were, and are, less ironical, which is quite wrong, about the double system of laws in which we live, which results in the fact that a case could be decided quite differently by a court in a member-State of the EU and a European court. That crumbling of the law, of the laws, of the States, which alters what democracies seemed to be tending toward – probably with no possible going back – should not sadden us in any particular way, but it should make us pay attention to what they never were and will never be, what Rousseau said they should be – without having many illusions himself – or what Mill treated them as if they were – regimes that, though not yet existant, were expected to become the norm everywhere.[1] The points on which democracies were expected to create new problems have been negotiated in various ways, not without experience of horrible accidents and crimes, which have been survived. But are they delivering, or have they ever given us, freedom, equality and brotherhood? What is happening now is something that is not unrelated to the democratic ideal, that is true, but that ideal does not exist and will not exist in the State that Hobbes, Montesquieu, Rousseau and Mill described. Some types of relations are being built among people that were not imaginable when the link of citizenship that tied each to the Sovereign was strong, or at least seemed to be so. We are not judging them because we do not yet know how they could evolve. We can only see a double outflanking of the States, one movement that seems to be rushing towards State structures, and the other which seems to subvert those State structures

[1]See the Preface we wrote to the French version to the *Considérations sur le gouvernement représentatif*, [16], p. 25, p. 46. Democracy has been described by Millin minute détail of its running, with its advantages in comparison with other regimes and its wories, even though he had no exemplaire of it before him.

and undermine them in an inverse way. Those two movements, which seem so far from one another, are not without analogies. This is what we would like to consider.

Our starting point will be three features that have constantly struck us since we started studying ethics, that is, more than a decade ago. The *first* is that, in a Lockean view, there is no more a Christian, Muslim, Jewish or atheist ethics – ethics being precisely an example of trying to solve conflicts or disagreements which arise because of the very diversity of the values at play in a situation – than there is a Christian, Muslim or Jewish State, or a State that would recommend atheism. Thus, ethics, like the structure of democratic States, cannot admit any religious or moral transcendency, since – without challenging or questioning that transcendency in individuals – which is often its subject and object, its real objective is to try and make women and men with very different convictions, and who will not change them, live together under the same rules. Disagreements, when they happen among human beings in the field of ethics, because of religious or nonreligious, moral or amoral, convictions, are solved thanks to methods which filter or compose those convictions, but do not side with any of them. The *second* is that the way ethical disagreements and conflicts are solved, when they are not brought to court, is quite like the way diplomats solve disagreements and conflicts among States, that is, following a strategy or a combination of strategies which, by definition, cannot be limited to calling upon the values of one or other party, in the case of a two-party conflict. The values of ethics are quite close to those of diplomacy, and cannot depend any more than they do on the particular or exclusive values of the State. The *third* feature confirms the previous ones in another manner. When one reads about each major topic of ethics – about the ethics of care, medical ethics, or bioethics in particular – one realizes that, whatever the language of the author, whatever his nationality, he does not say, in his own language, things that are really different from what someone else would say in their own language, and that he sees problems where they see them, and considers solutions in the same areas where they project them, exactly as books on diplomacy are not completely different in moving from one language to another, from one nationality to another. There is only a marginal difference – at least, a much smaller difference than one might expect from the beginning, even though languages order different ways of thinking[2] – in the content of a book on ethics written in English and one written in French, German or Romanian, provided, of course, that one takes into account the date when they where published. Admittedly, one could, like Hume, call upon some identity and permanence of human nature. However, we believe that the reason for the similarity is quite different, and is instead due to a design of ethics and diplomacy which is quite similar. That is what we intend to demonstrate now.

---

[2]Even in the names of illnesses. We have written an article on that point [4].

1.1 The Common Features of Medical Ethics and Diplomacy

There is no simple relation of analogy between ethics and diplomacy, between medical ethics and some form or other of diplomacy – since it is more a form of collaboration – but one can start with a few notable features that make them similar.

If one agrees to define diplomacy as the discussion and assertion of the interests of a State by one of its official representatives, in a language and a discourse that another State, or one of its official representatives, can understand, and, at least up to a point, accept, and if one completes that definition by inverting it, and speaking of the discourse of that other State, or of one of its representatives, as discourse that the State I belong to can understand and, up to a point, accept, if one thinks that this is one of the main features of diplomacy, then ethics – and medical ethics in particular – is similar to it, in a quite essential way. For what is at stake in ethics is having a discourse directed at someone who is likely to understand it or to listen to it, a discourse that tries to take the other into account so as to build a common platform which makes it possible to move forward at the same pace, or, if points of view are not reciprocally admitted, to continue nonetheless to talk to one another, to understand where the disagreement lies and to find a solution that makes it possible to transform it into a problem that can be solved together. The important common point here is that one does not have to share the values and the point of view of the other and impose them on oneself, but one should try to listen, to understand and to account for them in the other's language, with their own codes, as in one's language and codes, and to make them a theme of one's arguments. Like diplomacy, ethics does not disbar the emotions and passions which constitute its building blocks, but gives them a voice audible to the other, so that they will have a hold on them. The transformation of affects into interests that are submitted to discussion is an element that is common to both the two points of comparison.

Another major idea, which however, leads us towards appearances that will need to be corrected, is that, in the one case as in the other, the activity seems to be strictly limited by laws, by rules defined by politics. The diplomat defends the interests of his State and is not supposed to go outside the framework of negotiations that he was asked not to transgress. The patient cannot ask his doctor for something that the law would forbid, without his request being refused, and the doctor cannot afford to break a law that governs his occupation. We will see that that basic position is sometimes quite abstract, and that the fact that diplomacy and medicine stand at the limit of what is allowed and what is forbidden often implies, for the operations at stake to succeed, that there be some leeway that allows some latitude to be taken regarding laws. It is important though, that the diplomat, like the doctor, constantly feel that they have been commissioned by a State or a government which trusts that, in their positions, both will abide by the laws of their countries. Such trust is founded on the friendship and gratitude that a nation has for its doctors and diplomats. That trust however, is also related to several reasons that compel the State to adopt that attitude.

The first is undoubtedly that the doctor, like the diplomat, only deals with singular cases, that the law cannot foresee all the cases, and that the government having to enforce it must count on the loyalty of those who have to solve those singular cases, by finding, on their own, rules that may not always be applicable in other cases. Medicine, like diplomacy, remains singular, and the rules it uses to solve an issue or a conflict are more those of an art than those of a science. What is at issue is not always – and perhaps is only rarely – the application of the rules of a case to another one, though one may sometimes think, when it is possible, about changing the law and about transforming what was at first a case-by-case intervention into an application of it. That is not always the case, however, and one can hope that a State trusts its citizens enough to solve problems on their own, because they are in the best place to do so. The clever initiative of an actor, in a contingently produced situation, may be much more socially useful than the application of a law that has not provided for the case that must be solved, and thus places it on the Procustean bed.

The second reason is that the ethical – like the diplomatic – solution of a problem is often better, given the limited time-frame in which decisions must be made sometimes, than waiting for the law or the legal dispositions to be implemented before deliberating and deciding. It is even possible that problems can be resolved by ethics, whereas they would have left deep scars had they been settled by the courts. Some misogynistic attitude or religiously intolerant behaviour may sometimes be altered and improved more easily in ethical conditions than by the application of a rule which, if applied in a wholly uncompromising way, or even only quite rigorously, would be devastating in the mind of some parties in a situation. That does not mean, make good note of it, that ethics by its nature encourages cowardice, that it is but a dodging of the application of the laws, or that it makes it possible not to apply them or to avoid them, its only objective being to get a peaceful solution whatever the price of that peace. Admittedly, the aim of ethics, as of diplomacy, is truly to get to a solution that suits, if not everybody, at least the main actors in the situation – but that does not mean that there is no unacceptable position from an ethical or diplomatic point of view in some situations. No doubt the very emergency and necessity of a situation imposes a pressure to find solutions, but it is not obvious that those solutions need the agreement of all, or their unanimous assent. Consent, which is key to ethical agreements, does not require all the parties to an ethical situation to adopt the same values and conceive the situation in a same way. Though it does not consist in unanimity, it is the result of a weighing of points of view, the outcome of which may be that one point of view prevails over another, while the interests of the patient in a medical situation are secured and promoted as much as possible.

Consent does not require either that a State agrees with the values of another, but that it preserves the interests of the peoples or populations affected in case of state conflicts. Putting oneself in someone else's place does not imply adopting their values, but only understanding that they are theirs, that one allows them some space, that they understand one's own – which may be objections or obstacles to theirs – and that a composition which could be somewhat consistent, stable and sustainable is achieved. One may hate the values of those against whom one is in an ethical conflict, but one allows them to hold and express them, while they may give a more or less important place to one's reasons, motives or values, even though they do not share them at all and may do all they can to defeat them. In ethics as in diplomacy, one is working with the values of the other, not because one thinks they are better than one's own, or that they should become one's own, but to build a situation that will allow us to continue to operate, work or simply live together.

We are talking of values here – moral or religious values – of which we are saying that there can be none that are able to transcend all others, and we are asserting that it is in the immanence of negotiation that the transcendental conditions of a relative surpassing lie. We must nonetheless go even further, and also talk of the procedures and processes we use to deal with those values. For they do not escape mediation any more than the former, and are also negotiable. One does not have the right to demand that the other, if they do not go so far as to recognize one's values, should at least acknowledge the methods one uses to get agreements. Methods themselves are negotiable, for the modes of transformation of conflicts into problems are also diverse and need to be worked upon.

The third reason for the necessary trust in the doctor or the diplomat is that, even if I were to dream of a transcendental position, of an absolute point of view, of a sort of meta-structure that would allow me to go beyond the perspective of everybody, such a dream of universality and necessity would not go beyond the status of one position, one viewpoint, one structuring, among others – which I may have my own reasons to think is superior, but which may not appear so to others. This would be the case, even if I took it for a reality that could not be challenged in good faith. At most, what we are building together is a local surpassing, which includes the actors who are present, without resorting to any transcendental value whatsoever. It would be better to talk of a constitution of transcendental conditions but without asserting any feature of universality or necessity, though we may all be looking for them. The carer works at the social level, the diplomat at the international one, but they should make sure, if they want to be useful, not to reify what they do as if it could be generalized and used again immediately. The only thing that could be reused is the aim that was pursued by the weaving of an agreement, and which will, if fed with singular biographies, lead to other negotiated agreements in completely different conditions. In ethics, as in diplomacy, roles are never set once and for all: actors are not substances – they can be divided or aggregated in diverse ways. They constantly change as they become. The relations they try to strike up, to invent – perhaps to resolve difficulties, to unravel them – do not entail that the sort of beings between whom they are being made must be known in advance. Neither the roles of the actors, nor their limits, nor the format of the scene they create as they play it are set in advance. The text is to be invented. It is that very divisibility and multiplicity of actors which makes negotiations possible. What is being invented or to be invented is not without deep repercussions, which are sometimes beyond the actors who are present. When one says that each situation is particular, one does not mean that only the few individuals that seem to be involved in it are concerned. If the person is not the right unit of the processes one is considering, the individual is not either. Rather, through him, all sorts of social combinations and myths are being played out, which should be taken quite seriously, for they are real for those who promote them.

There are undoubtedly many other features that would allow one to compare ethics and diplomacy. Some of them could be easily deduced from those that have just been mentioned, while others would be new and have so far eluded us. The point on which we want to end our more or less articulated list of analogies is that one should never think that in ethics and diplomacy one always knows where the real debate, disagreement or conflict is, or that one knows how to make the difference between the fictitious, the evanescent, and the real, between what does not count and what 'really' does. Thinking that way would be a gross mistake. A.O. Hirschmann has rightly shown that as soon as one wants to take situations into account so as to provide a political or ethical solution, one should consider that interests should prevail somehow over passions, but that this does not prevent the latter from being taken into account quite seriously [6]. For us, because he did not take utilitarianism and its utility calculus into account, he did not go far enough down that path. The great advantage of utilitarian calculus, from its Benthamic origin, is that, in addition to the production and exchanges of what the actors consider to be real, it integrates into the very essence of the calculation – not through some *ad hoc* or fragile addition – a sort of second line which is as important as the first one, and which forces one to read the first differently. Thus, neither an ethicist – even though they loved arguments – nor a diplomat – even though they were very keen on defending the interests of the State they represent and the interests of peace – will lose sight of what Bentham called the 'axioms of mental pathology' [2] and what President Sadat before the Knesset called the psychology of adversaries. The models that are being used in ethics, as in diplomacy, could not have the structure of contracts which only refer to abstract, knowing and willing individuals. In the two activities whose similarities we have been looking for, infinitely more complex relations which are there much more mobile and unstable are in play. One should not however think they are not rational. We have just seen that mathematics – provided it is not as naively simple as in Rousseau, for example[3] [14]– could provide excellent frameworks in probability calculation and game theory.

1.2 A Few Objections to the Previous Set of Analogies

One must admit though, that despite the previous easily discovered analogies between the ethics of care and diplomacy, some objections lie on one's path and could become as many obstacles if not addressed. They may reveal themselves to belong more to the order of fantasies than of reality, but they should be tested if one wants to go further.

The first difficulty of too easy an assimilation of the task of the ethicist and of the diplomat is that the latter in principle works closely for the State he has been

[3]When what is at stake is to differentiate between *the general will* and *the will of all*, for example [14, p. 371].

delegated and commissioned by, and only thinks of its interests when he negotiates. In the ethics of care, the doctor is less constrained by very tight links with the State. He may, in the name of his ethics, treat an enemy of his country as well as a fellow citizen. Medicine without borders is, somehow, a pleonasm. Even though a doctor were treating a fellow citizen, he would quite often do the opposite of what the State would like him to do. That is why if, in order to cut back on jobs, a health minister may ask the manager of a psychiatry department to install a CCTV system on the premises, he may resist – and he even should – by refusing to consider as 'care' what is in fact only a security system. In order to protect a patient he may – and even should, and would be at fault if he does not – use a specific secret, the legitimacy of which most States admit, but which would make the other citizens accomplices to misappropriation if they used it. Thus there exists some relative withdrawal of the carer from the values of the State in the field of medical ethics. Though every doctor is linked by all sorts of relations to the State of which he is a citizen, every doctor is also a doctor of the world or a doctor without borders thanks to the oath he took when he became one. That is one of the reasons why the content of books of medical ethics so easily cross geographical and linguistic borders. Admittedly, some adjustments need to be made from one language to another, as from French into English, but when what looks like a distortion crops up, it is no more serious in ethics than in science, when it is acknowledged that one language favours the birth of a concept which might be better dealt with subsequently in a language different from the original, in which the changed understanding could not have been developed. The disagreements among languages in medical ethics are temporary and ethics becomes richer from those disagreements, provided one tries to understand them and account for them.[4]

In addition to a relative lack of interest of the caregiver in their own values and in some State values in the medical ethical relation, which is different from the interests of the State and its representatives in diplomacy, there is another difference which cannot but seem essential, and which may threaten our previous series of analogies. Even though the whole topography of the tricks that might be used in diplomacy might be mapped out,[5] which would render them less efficient – one cannot but see that there is no negotiation among States, or among commercial groups, or between employers and trade unions, which does not employ the whole gamut of such strategems: dividing the opposing side, wearing it down by tiring it out, linking unrelated issues to get concessions, using threats, being dishonest as to the real level

[4]This explains how to sign a « certificat de refus de soin » that French care givers fear and sometimes denounce so strongly has no other equivalent in English than 'to sign a waiver'. It is an interesting difference. It can be solved only once French-speaking people understand that a person who no longer wants any care is not necessarily refusing it to their care-giver. The play of personal implications in the English language does not operate in the same way in French, nor in any other language. The way ethics is formulated is obviously different from one language to the next, but it is not an obstacle to the fact that ethical issues are framed in roughly the same terms by English and French speakers.

[5]That endeavour tempted rather than it was attempted.

of satisfaction or dissatisfaction at which one would leave the negotiating table, organizing an under-diplomacy intended to prevail over that which is more apparent, demanding that negotiations be short, or on the contrary extending in their duration, and so on and so forth. If tricks are used by the carer to get their patient to behave in some way – to accept treatment, or not to refuse the protocol best adapted to their case, etc. – if, since the Greek Antiquity one has known that there is no care without the art of convincing the patient of the legitimacy of the care that may be painful, uncomfortable, long, or disabling,[6] and if some lies may win the game or give more hope than the reality of the situation does, it is nevertheless rare and frankly incredible for the relation between doctor and patient be founded on lies. Some psychiatric treatments are partially founded on some sorts of bargaining with the patient, but the ultimate goal cannot be to deceive them, if only because that game would be short, could not be repeated, and would very soon have the opposite effect to that intended. Trust is an important element in the medical relation and, if it is not destroyed by betrayals and lies, even when employed for a good reason, it is in any case weakened – even though the patient retains some regard for the technician that his doctor is.

However, one should take into account a much more important argument to divide ethics and diplomacy. Is it not very strange to use the position of a diplomat to characterize medical ethics, even though the job of diplomat has its own particular ethics and deontology? Is there not thus a difficulty in comparing the values of medical ethics with that of diplomatic practice, because diplomacy itself has its own values which are distinct from those of medicine? Do we not suddenly make a shift from the consideration of values to that of facts, or from facts to that of values, thus committing a fallacy that some thinkers like E. Husserl ([7], Ch. III–VIII, p. 55–211) or G.E. Moore ([9], p. 62–73) denounced – albeit in somewhat misleading way – as a *naturalistic fallacy*? Even though one might have preferred speaking of the 'factualist' fallacy, the importance lies in the argument rather than the terminology. One would have mistaken a job, with its technicity, the technicity of its procedures, for the values of the ethics that characterize and model that technicity, without being the latter. One may hold one's position well or badly – it is not the fact of the position which is good or bad. Thus saying that ethics is *a form of diplomacy* would only seem to have more meaning than saying it is a form of medicine, or of legal practice, or of judicature. This is precisely one of the questions posed in *Gorgias*: is rhetoric good in itself? Or can it be good or bad? Is there some good or bad ethics?[7]

This is not a trivial argument, but it is obvious at once that it is founded on the supposedly evident distinctiveness of the value and the referring to it by a certain number of acts. There is no certainty that it is easy to establish this distinctiveness,

---

[6]Plato underlines that point in *Gorgias* [12], 459a-460b, p. 298–303. Gorgias's brother was a doctor, and Gorgias claimed that he had on many occasions managed to convince patients to accept a treatment more successfully than his brother.

[7]One will not mention the argument that by assimilating ethics and diplomacy one wrongly assimilates ethics to the ethics of a profession, as if that profession were the real driving force of it. Why should it be that one and not some other?

and thus to separate the value from the acts it is supposed to measure. When one measures something, it is always difficult to remember that one's measurement is linked to the instruments and methods one uses to make it, that it only has meaning in relation to them, and that the projection on to the object of a true measurement to which one would adapt gradually, by trial and error, is a pure fantasy. Pascal showed what he called the 'absurd humility' which consists in believing that one knows in what the values of 'reason' and 'justice' consist, and that only the inadequacy of one's instruments is responsible for preventing one from expressing those values in all their splendour.[8] One should go even further in the criticism of that Platonism which sets the difference between value and fact, which treats value as if one knew how it measured the fact, although value never appears but as a sort of trick in the indefinite play and counterplay of measures. It is by refuting Platonism that one will understand first, why it is not wrong to assimilate ethics to a sort of diplomacy, and, second, and even more concretely, how diplomacy and ethics work together in a game that reduces the weight of the State and that of the laws that it passes.

1.3 That Those Objections Could Be Countered by Saying That They Weaken One Another, and That on the Contrary, the Answer Will Allow Us to Explain a Contemporary Movement

Nowadays diplomacy seems to have arrived at the confluence of two demands which, even had it tried, it could not have escaped. The *first* is that it has become permanent, which is to say that it cannot be satisfied with the resolution of conflicts on a case-by-case basis, that it has become necessary to prepare for conflicts, perhaps to take advantage of them, but above all to obviate them even before they are kindled and become destructive. That *permanence*, which dates back to the seventeenth century, and which has but reinforced itself since then, has led to the formation of a body of diplomats, each familiar with the issues and language of the others. A sort of network has been built which has been often described and which allows international problems to be better known and better handled by those professionals than by the governments which only hold their positions for short periods of time, and which in reality depend on specialists who are more competent than they are. This is *the second feature* that characterizes contemporary diplomacy. Far from being a simple representative of his government, a diplomat nowadays participates in governance

[8]*Pensées*, Sellier, 67: 'On agit [...] comme si chacun savait certainement où est la raison et la justice. On ne trouve déçu à toute heure, et par une plaisante humilité on croit que c'est sa faute et non pas celle de l'art qu'on se vante toujours d'avoir'. ([10], p. 852). [«We behave (...) as if everyone knew for certain where reason and justice lie. We are constantly disappointed and an absurd humility makes us blame ourselves and not the skill we always boast of having » (*Pensées*, 33, [11], p. 7)].

by transforming representation into a very fragile mask. Representing a government in a negotiation with other States unavoidably means setting in motion their relations, and substituting to the simple representation, which could not remain passive, acts which, being constitutive of politics, are as real and even more real than those which could be performed by governments. Though one cannot do anything without representations, one cannot represent without doing anything. The leeway that is needed to represent is so important that it amounts to an act in itself, which, imitating nothing and being in thrall to nothing, has its own autonomy, its own construction. There is a dialectic game, a Galilean game, one could say, which results in a movement never being in itself what it is without being what it is, from the point of view of other movements, indefinitely, that is, with no absolute point of view that would stop the process.

The same movement exists in ethics. Admittedly, over about the past half-century, and up to today, ethics has become important to governments virtually everywhere in the States which have had the means to develop it or to let it develop itself. Once it has been established, it is clear that its function of counselling governments[9] has allowed it to become increasingly demanding. If governments are reluctant to take a measure because they fear the reaction or the resistance of a lobby, of a church, or of a certain type of morals, health professionals, who are in contact with patients and families, who know the problems and mindsets which confront them, are, for their part, in the best situation to say what should be done, in such and such domain, even though they were divided amongst themselves for moral, religious or ideological reasons. An extremely powerful principle of reality – that of their job, their care – substitutes for their ideological disagreements, which, admittedly, should not be set aside, but should not monopolize all the attention either.

To consider only these two movements, they are making a pincer attack on the State, from above and from below, as it were: on the one hand, in the sense of the weaving with the other States that is undertaken through people who necessarily go beyond their role as representatives; on the other hand, in the sense of a pressure for laws to be drafted which could only be drafted with the help of multiple committees and councils which the State itself has created and which, on the whole, order it to do what it should. Nobody complains about it, but the old conception – one should rather talk of the old fantasy – of democracy is being countered, which gives the impression that democracies are declining, which is strangely in contradiction with a powerful ethical movement, which nobody tries to question any more. Truly, it is not a question of a decline of democracies which have only ever existed in the heads of ideologists and a few political philosophers. It is a powerful movement in which the State's prerogatives are collapsing behind a façade which seems to preserve some of its authority. Though dangerous, the nationalist votes – which benefit those who hate freedom and a multiplicity of values, and indicate a desire to cling to values that have

[9]When it is about the CCNE (Conseil consultatif national d'éthique, the French National Advisory Commission on Ethics), for example.

no chance of ever existing – are but the symptom of a fear of the game through which movements overcome one another by fantasizing the end of that transcendent overcoming movement, instead of understanding how the game works and is regulated, and getting a real grip on it.

Today, there is – in quite a contemplative, but also quite active, way – an *inversion* of issues as they were set a few centuries ago by the supporters of the 'contract' to represent politics. A reverse problem cannot be solved the same way as a direct problem, and the solution of the former is often more fruitful than that of the latter. Starting from conscious and willing subjects, it was wondered how everyone could consent to submit to rules, laws and authorities which they had not deemed good, and had not wanted themselves where a majority had decided them against them. Manifold imaginable variations have been deployed by a wide diversity of authors to try and solve that problem, which is insoluble when posed in those terms. These authors tried to deduce the relations men should have with each other from fictitious subjective substances. Today, it is the opposite path that seems to be the rule in the two fields we have brought together. What is now favoured, in line with Hume and Bentham, who rejected the contractualist model to speak of States, is the relations of the substances and starting from those relations rather than from supposed substances. Relations between States have more reality than States themselves, and the rules that govern the relations among citizens cannot always have the rigidity of laws, even though laws are part of the situation and of its solution. The constitution of ethics is built on a relation in crisis which demands resolution by the giving itself objects, subjects, a format, limits, procedures, resolution processes, by imagining their future and trying to operate on it. Relations come before the beings they place in the presence of one another, and they mark the rhythm of their constant becoming. Though ethics generally respects the legislation of the States in which it develops, it does not, in itself, have any predetermined framework, and it determines its own frameworks.[10] It is the necessity to find a solution that leads to the creation of the situation. It does not exist before the recognition of that necessity by subjects who should exist in themselves and would get in contact on that basis. If morals can be fundamentally substantialist, ethics, like diplomacy, starts from the disagreement or the conflict, can only be event-related, and, far from considering its subjects and objects and their futures as given, produces and deduces them, without realistic prejudice, and consequently, without preliminary rules. The rule itself is conceived only as a temporary balancing, which is necessarily unstable, between the parties it puts in presence, and even those it will put in presence.

Will it be a surprise then, that if morals are exploded by an ethics which seems to perform well only in its plurality – because it is when morals are the most diverse and divided that ethics is most required and justified – diplomacy today is quite divided itself? Far from representing only States, diplomats can now act in very diverse capacities: NGOs, churches, organizations which consider the environment on a global scale, multinational companies, which do not depend on the authority of the

[10] That is why it is so easy to read the systems of ethics written by foreigners and almost instantly use them.

States at all, and would be suspected by State diplomats, can weigh on the inter-State decisions, by imposing their own standards, as if they were those of Reality itself.[11] In some ways, it is true that they have become Reality itself, while the State is no more than a pale fiction, claiming an unreal law and having little grip on a reality that escapes it on all sides.

Should we be alarmed by this result, or is it the result of a necessary process which has nothing to do with the frameworks within which some would like to keep it? A process is at play that mocks the States, and only uses them and their policy to utter a speech that is no longer theirs, and which has probably only ever been intimately theirs in the imagination of some philosophers. After all, ethics has known for a long time that its work is not to keep the patient within the rules, that illnesses can be the occasion for what Nietzsche called the 'great health', which uses elements of deficiency to form a speech that has no longer anything to do with withdrawal and tightening.

What is particularly striking is the coincidence between the unbelievable success of ethics, which considers the 'medical' and care in all their aspects, and the explosion of diplomacies, which has occurred, like the former, in the last half-century. Such a coincidence is so strange that one might wonder if the two phenomena have a common root. That is the question that we would like to study now.

1.4 The Evidence of a Common Root of Medical Ethics and of a Diplomacy That Outflanks States

First, one type of arguments must be refuted – though with some qualifications – in favour of an alternative. The *first* aims to refer these two phenomena to a human nature which can be found sometimes above and sometimes below State level, while negating their factual, and, for us, quite new, character. If 'human nature' is understood to mean some permanent characteristics that endure across the geographical and historical dispersion of certain types of living beings, then the mention of such a basis for reasoning makes one quite skeptical, as one rather believes in the constant folding and unfolding – which one has called Galilean – of movements in a dance of reciprocal influence with other movements, without pause or stop. If that unfolding follows any order, then that only is what should be accorded the name of 'nature', though quite reluctantly. There is however, a *second* type of argument

[11] In *Diplomacy and the Making of World Politics*, [13, p. 4], the authors of the Introduction (O.J. Sending, V. Pouliot, I.B. Neumann) start an uncompleted list to show how the framework of States has become too narrow for diplomacy: there are paradiplomacies, diplomacies of small States, non-governmental organizations, business diplomacy, public diplomacy, bilateral diplomacy, multilateral diplomacy, poly-diplomacy, catalytic diplomacy, star diplomacy, real-time diplomacy, and triangular diplomacy, so that, today, diplomacy has become 'A claim to represent a given polity to the outside world', as the States are no more than partners, not necessarily preferred interlocutors. Was there ever a period of time when they were exclusively so?

which seems to bring more evidence of what one is looking for: the establishment of doctors without borders – whose usefulness is indisputable – is sometimes accompanied by organizations which seek to address, or are invited to, the negotiation tables around which representatives of States talk, in order to indicate places where humanitarian disasters should be prevented (like hunger, complete absence of education, wars, the stranglehold of zealous religious figures on populations to the detriment of other religions or irreligion, etc.), and also to push governments, which are normally quite opposed to that sort of endeavour, to launch armed actions in the name of a contested right to intervene. Those who have worked most in that sort of diplomacy are quite disappointed and bitter now, not because of what they have done, but because of what States have done as a result. We think that their bitterness is not entirely well directed, and that they could equally well lash out at the equivocal character of their own action, which consists in deliberately making use of States for an end that is not theirs.

One could wonder if, now, as diplomatic authorities are extraordinarily divided, and new forms of diplomacy have appeared, there is the equivalent of that plurality in ethics. It seems that the type of conflicts that lead to ethical debates are increasingly linked to the crumbling of the State which results from the relative facility with which secular Republics can be attacked. A State that is weakening is also a State that is more or less temporarily supplanted by religious values and authorities, which were thought for a few centuries, in some countries at least, to be reduced to a symbolic activity, lacking any possibility of giving life to an imaginary supremacy. One of the greatest merits of the State, even as a fantasy and ideology, has been to ensure, in the main democracies, and, more broadly, in authentic republics, that some religions, some philosophies, some viewpoints on the world and others, would not reign supreme. The collapse of the State, even as a myth, is not without formidable consequences. It may be that any call for the strengthening of the State is too late, in that it is no longer possible. Conversely, it is still possible, in the great number of schools which still practice republican values, to teach at a very young age, at the youngest possible age, the values and behaviours that Bacon and Locke recommended, though in their case only for the upper echelons of society.

Every child whatever their social origin – not only those who intend (or are intended) to become leaders, diplomats or doctors – should not be given lessons in morality or catechism – which are not for schools to give – but should be taught to express their wishes, their will, and their interests, in the language of the other, while those others would be doing the same for their own wishes, will, and interests. That teaching does not exactly consist in love of the other, nor even sympathy – although it does not exclude them. It does consist in teaching some mode of calculation, of some logic, of a propensity to argue, through which all values and social behaviours are transmitted.[12] This is how some sort of ethical permanence could be

[12]The form that such teaching takes in Locke is that, from his youngest age the child knows where he is situated socially, and will behave according to that accurate representation. On this point see *Some Thoughts Concerning Education*, ([8], pp. 73–8). Bacon ([1], p. 56–8) had already

obtained, which would be the mirror-image of that permanence which diplomacy realized it so much needed as early as the first half of the seventeenth century.[13] It might be objected that such a conception of otherness is too narrow, or too analogical, and that it is precisely too ethical to be really ethical. References might be made to a phenomenology of otherness that claims to situate us much more archaically than the interlocking or the concatenation of what I think the other is going to do, and what the other thinks I am going to do – the principle of which can be found in *Moral Thinking* by R.M. Hare [5], which denounces the simplification and the diversion that such a conception of the other, which is infinitely too detached, instrumental and calculating, is.[14] We are only suggesting here that we are not to be deceived by a supposed archaism of otherness, that we cannot see why it should be more interesting to discover the existence of the other through the experience of shame, as, for example, in *Being and Nothingness* by Sartre [15], who uncovers that existence for us through that affect, without explaining himself much on that starting point except by paying tribute, without saying it, to a sort of biblical memory which makes the gaze of the other the occasion of a fall and degradation in which Adam and Eve recognize that they are naked. As a fundamental figure of the other we prefer, for our part, that half-passive, half-active making of a commensurability with the other which allows us to form a society with him, and which makes us constantly conscious not of transcendental values, but on the contrary of an indefinite relativity. Let us not think that this is difficult or artificial: the very operation of passions, according to a double association – of ideas on the one hand, and of pleasures, pains and sensations on the other – points to the place of the decisive articulation of utility calculi.[15] The gap, reflection and distance are already sketched out by the very structure of passion: what is needed is only to make that articulation more active and informed by teaching children to get hold of it and practice it.

1.5 Conclusions

1. The reader will understand that if I have deployed a humanitarian argument to illustrate the hypothesis of resemblance between medical ethics and diplomacy, my aim was rather to point to a symptom than to join forces with humanitarianism, which I wish to avoid for a reason that Pascal gave, which is that moral conceptions of the world are very dangerous: 'Jamais on ne fait le mal si

---

highlighted, through the idea of travelling, the art of culturally – that is geographically, ethnologically and historically – situating oneself.

[13]Richelieu was perhaps the first to be acutely conscious of it.

[14]See the chapter we have already written on the ethics of R.M. Hare in the first volume of *Rethinking Medical Ethics* ([3], p. 81–138).

[15]We have already mentioned this point: see p. 7 above.

pleinement et si gaiement que quand on le fait par conscience.'[16] Evil, when done gaily – for it is not always possible to say that war is good, even when undertaken for noble reasons – is often linked to some anger against the States which are accused of doing nothing to prevent massacres and famines. That anger makes the persons who promote such humanitarianism so politically unstable that they are seen to cross the entire political spectrum in the blink of an eye. This can create problems for the ideological stability of those who promote such humanitarianism: it can be almost comic to see spokespersons of humanitarianism veer from a very moralistic leftism to a security-fixated right posture when governments change.[17] One might wonder whether humanitarianism is not sometimes connected to some contempt for humanism. Undoubtedly, confusion of these two attitudes is to be avoided.

2. The space of some bypassing of the collapsing States that we have described has given us the opportunity to show some irony towards nationalisms which, though they seem to be expanding, are rather for us the sign of a discursive inflation which can have no grip on reality. On the contrary, having left aside economic issues, we have not talked at all of the action of multinational companies before whom a huge field is open, which they might have opened for themselves to favour the imposition of an order, or a new order, as it were, and which they intend to substitute for that of States. It would be especially interesting, from that point of view, to look more closely at how the pharmaceutical industry – or that of medical technology – works at world level. It is a pity that ethics does not study these problems more intensively. It may not be by accident that the pharmaceutical field has been a blind spot in ethics for a long time, for as ethics escapes a tight relation with States, it is in the interest of the pharmaceutical industry radically to free itself from its control. It may pretend that it is in tune with the very movement of ethics, while simply taking advantage of a system of relative deregulation. Care, being more framed by the State, is more visible for today's ethics than pharmaceutical treatment, which, although constituting in practice the other half of care, is less under State control.
3. In my view, we should avoid talking about a 'decline' of democracy. The problem is at the same time less serious and more serious than that, but it is in reality something else. Talking about 'decline' implies making a reference to a democracy which, in reality, has never existed and cannot exist, by the explicit admission of those who, in the eighteenth century, seemed to recommend it. There is no pure subject thinking in itself, which is being reduced to a pure intelligence of the common good and to a pure will to pursue it. The model

[16]«We never do evil so fully and cheerfully as when we do it out of conscience» (*Pensées*, § 813, [11], p. 245).

[17]One could object that these are a small minority of people who commit only themselves, without questioning the movements of the doctors involved in humanitarian endeavours. This is indisputable, but the idea of humanitarianism itself implies an attitude of relative contempt towards State structures, since what is at stake is to put pressure on them, or to bypass them: in short, to use them for ends that are not those of the States.

according to which the general will, the only authentically democratic will, is to be deduced by opposition to the will of all could never be anything but a contemplative one, and one incapable of producing any particular action. The law is never anything else but the result of a confrontation of real wills, with real interests that they seek to promote, even though its temporality is not exactly coincident with them. A free discussion of and about those interests, whatever their nature, by the involved individuals, provided all can participate and vote on it without being troubled about what he has voted for, is enough for democracy to exist. Diplomacy can never aspire to be more than an approach adopted simply in the service of nation-States. It is in those conditions that ethics develops, and we have already noticed that it was only when moral and religious values were widely pluralistic and contested that ethics came to mean something. When the State imposes on citizens that which they must believe and practice in matters of morals and religion, it is still possible to talk of a republic, but not of a democracy. However, although ethics can only be born in democracies, it can be exported afterwards to many other regimes, and can produce subtle research, notwithstanding the fact that it could never have been born elsewhere than in a regime that knows and accepts plurality. The Lockean State is of course more likely to *kindle* ethics than the Hobbesian State. However, ethics can acquire its meaning and its role in the latter, under a theoretical or symbolical form, and develop once it has been born in the former. Once again, however, all that we have said was to demonstrate that there is no purely Lockean or Hobbesian State, and that they are only trends or characteristics which have no reality.

4. To these three conclusions one more must be added. It is not common to present medical ethics as a sort of diplomacy. What prevents this move is the ideology of the person, which does not allow the posing of ethical problems in terms of mediation because it moralizes its demands and considers them to be absolute. Diplomacy is bound to be more relative, otherwise it would go nowhere and could only constantly let conflicts and wars become worse. By fleeing those absolutisms, but also by recommending that close attention be paid to the particularity of each situation, we have to equip ourselves with another instrument to support the doctor in all his diagnoses and prognoses: the calculation of the probable, which belongs to a specific rationality upon which ethics should draw, unlike morals, the law or laws of which do not require it. We will analyze this major topic and tool – the ignorance of which results in huge and avoidable mistakes – by considering it in the favoured field of psychiatry, where it is radically unavoidable. We will extend it to all the sectors of medical ethics, which will allow us to distance ourselves from dangerous principles like the precautionary principle, which may seem superficially rational but which cannot succeed in the attempt to substitute morals for ethics.
5. Finally, the Lambert case in France, which has just ended with the death of Vincent Lambert, opportunely provides us with an argument that confirms our idea that ethics is a diplomacy, since it highlights an element that we have hitherto left in the background. When law and ethics are mixed, or when the law interferes with cases that ethics could and should have resolved, it only sows contradictions

and then brutally solves them. When the press intervenes in ethical issues by relaying them, as it were, with loudspeakers, those issues forfeit any chance they had of being solved. Are diplomatic affairs conducted in public? They do not lack rationality, and the reasons why one option is chosen rather than another can be revealed, but there is a time to do it, and it is necessary that the actors directly concerned should be the ones who meet who settle the issue. As soon as lobbies interfere and raise their voice so that others can hear and give their opinion on issues the specific circumstances of which they do not know, intelligence and discernment, which could have got everyone to agree, are doomed to be prevailed over by arbitrariness and force.

Bibliography

1. Bacon, F. (2006). In M. Kiernan (Ed.), *The essayes or counsels, civill and morall*. Oxford: Clarendon Press. Particularly, Essay XVIII, Of Travaile.
2. Bentham, J. (1843). The Axiomes of pathology. In *Pannomial fragments, the works of Jeremy Bentham*. Edinburgh: W. Tait.
3. Cléro, J. P. (2018). *Rethinking medical ethics*. Stuttgart: Ibidem.
4. Cléro, J. P. (2018). Ethics and the increasingly English-speaking psychiatric tower of babel. In *Annals of the University of Bucharest* (Philosophy series, Vol. LXVII) (Vol. 2, pp. 3–20). Bucharest: University of Bucharest.
5. Hare, R. M. (1981). *Moral thinking. Its levels, method, and point*. Oxford: Clarendon Press.
6. Hirschman, A. O. (1977). *The passions and the interests: Political arguments for capitalism before its triumph*. Princeton: Princeton University Press.
7. Husserl, E. (1959). *Recherches Logiques*, T. I: *Prolégomènes à la logique pure*. Paris: PUF.
8. Locke, J. (2010). *Some thoughts concerning education*. Cambridge/London: Cambridge University Press. See in particular §94.
9. Moore, G. E. (1993). *Principia ethica*, revised edition by Th. Baldwin. Cambridge: Cambridge University Press.
10. Pascal, B. (2004). *Les Provinciales, Pensées et opuscules divers* (p. 852). Paris: La Pochothèque, Le Livre de Poche/Classiques Garnier.
11. Pascal, B. (1995). *Pensées, 33*. London: Penguin Books.
12. Plato. (1991). *Gorgias*, in Plato, Lysis. Symposium.Gorgias, Cambridge, MA/London: The Loeb Classical Library, Harvard University Press.
13. Pouliot, V. (2015). In O. J. Sending, V. Pouliot, & I. B. Neumann (Eds.), *Diplomacy and the making of World politics*. Cambridge: Cambridge University Press.
14. Rousseau J.-J. (1964). Oeuvres complètes, III, Du Contrat social. Écrits politiques, NRF Gallimard.
15. Sartre J.-P. (1984). *Being and nothingness* (trans. Barnes, H.E.). London: Routledge. [*L'Être et le Néant*,1943, éd. Gallimard, Paris].
16. Stuart-Mill, J. (2014). *Considérations sur le gouvernement représentatif*. Paris: Hermann.

Chapter 2
Ethics of Risk Taking in Psychiatry the Game of Risk and Probability

Abstract There is no medical situation that could escape risk taking; however the main characteristic of psychiatry is that the risk of a treatment or of a release from hospital -after involuntary commitment- is not only taken for the patient (as in other medical sectors); it is also taken with a view to the close contacts of the patient and the other people he will be able to interact with. We advocate for a medicine that may enjoy room to manoeuvre in order for the doctor -particularly the psychiatrist- to make the best decisions for the patient; but it is evident that a decision taking must not be an adventure. What is a rational decision? We will mark our Bayesian preferences.

Keywords Adventure · Autonomy · Bayes · Bayesianism · Calculus · Danger · Person · Probability · Prudence · Risk · Suicide · Trace

> 'If you took the maximin principle seriously then you could not ever cross a street (after all, you might be hit by a car); you could never drive over a bridge (after all, it might collapse); you could never get married (after all, it might end in a disaster); etc. If anybody really acted this way he would soon end up in a mental institution'. Harsanyi J., [7], p. 40.
>
> 'One of the origins of the theory of fictions was Bentham's outrage and revolt at the existence of many absurd and immoral fictions which cluttered the English legislation and law and were considered to be essential to court decisions in some cases.' Bouveresse J., ([3], p. 90).

Psychiatric medicine raises particular ethical issues linked to the specificity of the illnesses and patients it treats and the nature of treatments it provides. Its differences with the ethics of the other sectors in medicine was noticed a long time ago. The mentally ill patient worries people around him more and in different ways than other patients, for, as Pascal said using a good image, when somebody limps, he knows very well, and people around him know, that it is he who limps, whereas in mental illnesses, he who is supposed to be ill may accuse us of being mad and not judging

J.-P. Cléro, *Reflections on Medical Ethics*, Philosophy and Medicine 138,
https://doi.org/10.1007/978-3-030-65233-3_2

properly.[1] Thus one may fear that the patient does not have the same judgment of the value of life as us and that he may be more likely than other patients to commit suicide, and that, if potential aggressiveness is not directed against himself it may be directed against others – his family, his broader circle, carers, and the other patients if he is in a hospital. Such concern, whether well or badly founded, is increased by the fact that, when one is not trained to understand how the patient considers his own life, the latter necessarily escapes, and even when one is trained, he may also escape. As a result, even though the medical relation always occurs in a legal context, the pressure of the law is stronger in psychiatric medicine than in other medical sectors. Thus people can be locked up against their will, and that locking up can be considered as an element of the treatment by constraining the patient to that type of confinement, and thus going against the most explicit wishes of the patient. Further, such legal confinement deeply and unilaterally changes, for that patient, both the balance of rights and duties which usually holds good for the other members of society, and, quite as radically, shifts the boundary between his private and his public life (the communications of the patient with his family and friends can be important elements of treatment and management during his illness, and belong to his medical file). It is not a question of punishing the patient for an imaginary mistake he would otherwise have made, it is not a question of humiliating him, but the categories we are so much in the habit of using to think about ethics, which are of a nature increasingly sanctioned in the law – which refers to the decision-making autonomy that we have been told for one or two decades belongs in the end to the patient only, and to the series of concepts that result from it, such as respect for the person, for the personality,[2] for dignity, and for consent, enlightened consent – are in quite a weak position from which to solve many of the ethical problems in psychiatry. It is not true that, in psychiatry, the autonomy of the patient is always respected, simply because that is not possible and the situation does not allow it. Attacks against the categories in which one wants to frame, or which one even thinks have to frame, ethics, even in contexts where they seem least well-adapted, are parried in a well-known manner. The patient is treated as if he had wanted – or would have wanted – what is done to him, if he had not been ill, and had thought of himself when he was still in good health, if he had been in good health, as a patient suffering from the illness he suffers from now. In the name of his autonomy, he would have consented to being deprived of it. In the name of his freedom, he would have admitted that another or others, deemed more rational than him, should have at their disposal during the time he is provided care, the freedom to treat him. In other

[1] *Pensées*, 98: «How is it that a lame man does not annoy us while a lame mind does? Because a lame man recognizes that we are walking straight, while a lame mind says that it is we who are limping. But for that we should feel sorry rather than angry» ([19], p. 25).

[2] The two concepts of *person* and *personality* are differentiated in Kant's works in the following manner: 'A person is a subject whose actions can be imputed to him. Moral personality is therefore nothing other than the freedom of a rational being under moral laws.' ([10], p. 378). From this it follows that 'a person is subject to no other laws than those he gives to himself (either alone or at least along with others)'.

words, in order for ethics to maintain its system of values, frankly fictitious imaginings need to be conceived, which are so contradictory, so unbelievable, so inconsistent, and so distorted that those fictions are pure 'fallacies', that is, stories that are being told, but which are radically false and deprived of any relation to reality. However, really to innovate any further than by building precarious balances or by perching old notions on stilts, it is necessary to divide ethics according to other categories which are closer to utilitarianism, an approach more willing to know the *individual* and the *trace* that the individual leaves, than the *person*.[3] All these categories are more closely related to utility, that is to the play of pleasures and pains of that individual and the community to which he belongs, or to the preferences of the individual and the well-being of the community, than they are to the impossible autonomy of the person; and they all invite consideration of risk and probability.

It is precisely these concepts of *risk* and *probability* that must be at the foundation of a triplicity (individual and trace/preference and well-being/risk-taking and probability) which seems both less rhetorical than the ethics of the person, and much better adapted to the needs of psychiatry. We have not mentioned risk as a relation intimately bound to the probable in the first chapter, though it is one of the major categories of all the fields of medicine, and especially so in psychiatry, in that it is more salient there than anywhere else in care, and, above all, in that the risks are of a different nature than arise elsewhere. Admittedly, any medicine presents a risk for the carer, but, in general, the doctor does not bear the risks of an operation, of a treatment, of letting the patient go home or go back to work too early. Instead, the patient deliberately, loyally and knowingly assumes at least the lion's share of that risk, with more or less help from his family or persons he trusts. In psychiatry, there is a greater risk for the carer in the medical decision, because of nature of the patient concerned. First, the help that the patient can offer, when it is not non-existent, cannot be considered as really and clearly contractual and mutually agreed as in physiological medicine. Second, the doctor is not necessarily helped in his endeavours by the law either, and maybe even less so by the general public which, quite legitimately, asks for security from the authorities that are in no position to ensure it absolutely, but which has become increasingly habituated to receiving security instructions. This may have led to some excess in the application of the precautionary principle which, on the pretext of prudence and public safety, may prevent any action bearing the slightest hazard by declaring it bold and daring, or even rash or dangerous.

Is it not possible to develop and define rational attitudes towards risk which do not depend solely on the universalization of the maxim of action? Is it so difficult to determine some of them for patients about whom it is often thought that a feature or effect of their illness involves or leads to the fact that they do not act rationally, that is, in a way that can be accepted by the whole community, or at least by the major

[3]We tried to demonstrate its weaknesses in our first volume: see *Rethinking Medical Ethics*, Ibidem-Verlag, Stuttgart, 2018, pp. 35–61.

part of it? A 'rational' behaviour here means a behaviour with an efficient strategy, and also an attitude and behaviour towards the risky and the probable which might be called *right* or *wrong*, as English speaking people say, who have more words at their disposal to differentiate *moral good* < *right* > and *moral evil* < *wrong*>, from *good* and *bad* in the sense of *pleasantness* or *unpleasantness*. We will wonder whether – in order for the attitude of the doctor in charge to be more open towards risk, and for him to give a more reasonable and equitable role to the probable – it will be necessary to change the rules of the law in that respect, and, even more, to educate the public, which does not always know –two centuries after Laplace and Poisson posed the problem – what to do in a probable situation.

2.1 The Appearance of Risk in Psychiatry and Its Forms

The *first* risk – the most tragic one, or at least the one which is considered to be so – is that of the possible suicide of the patient, which must be prevented, probably more than in other medical sectors, and with another intention. Certainly, great sufferings and distresses can arise from all sorts of illnesses in all the hospital departments, and the doctor may be asked by the patient to put an end to his days, but the doctor is, most often, relatively protected against a formidable and direct responsibility by the idea that if the patient really wanted to end his life, he would not need the doctor's help, which would be criminal in the eyes of the legislator. A doctor does not have to help his patient to die. He can now only help him to suffer less, or at least not any more. The portion of responsibility that the patient is attributed, maybe by fantasy – in any case, the law itself has him assume it so that he takes the risks and not only runs them – relieves the doctor of a part of his. If the issue of suicide is more delicate in psychiatry than elsewhere, it is probably because of the very intellectual idea that if the patient committed suicide, he would not know why he did it. He would not have 'good reasons' for doing so, since the reasons he could have given – even though they would have been valid if given by someone else – would be reasons which he might have rejected before his illness, which has vitiated his judgements, his will and his decisions. Thus, as has been said about the art of governing, what is at stake, for a leader as well as for a psychiatrist, is protecting people against themselves, defending, even though the individual himself rejects it, the 'good side' of the person, which that individual cannot defend himself, against the 'bad side'. However, should medicine exercise the function of upholding one part of a person against another part? Should it, out of principle, discredit or try to counteract and destabilize a wish to commit suicide by treating that wish alone as something which demands cure? The doctor is driven to assume the mask of the person, which encourages the individual to raise it against himself, and oppose it to his current wishes as if they were those of the 'false me', which promotes its values against those of the 'true me' – of which the doctor poses as the custodian and only guarantor – which forbids them. The advice that the consultation offers quickly becomes a requirement to conform to a principle of reality of which the doctor is the sole arbiter, in accordance with the law.

The *second* risk is that which the doctor runs when he does not apply detention and restraint measures in relation to a patient likely to be a danger not only to himself but to others as well. Once again, the concept of the *person* is convenient for its ductility in justifying what is actually its direct opposite, because the person, which is different from things, is precisely that which in us freely sets the standards, the rules of our lives. In the name of that freedom, by some strange oxymoron, the individual who is suspected of being dangerous is to be deprived of the freedom of going where he wants, of refusing treatment if he wants, for it is possible to think – at least one gives oneself the right to think – that, had he been aware of the hazard he presented, he would have refused the freedom he is asking for himself. Since the illness keeps the patient from accessing that awareness, the carer must take his place and do what the patient would have done in his stead. Fundamentally, even though he says that he wants to go out, for example, one can think that, in another way which, admittedly, lacks empirical reality, he still really wants to be detained for as long as it takes for him to get better, that is as long as he remains a danger to others, even if no dangerous act has yet been committed. At least that is what he would want if he were in a state to assess the situation soundly, and what he will have wanted if one day he is cured, and carries himself in thought to those disagreeable events of his past.

Moreover, and this will be the *third type of risk* that we will study, it is not possible to accept some acts socially, which, without being criminal, nonetheless threaten the social and economic life of other individuals, who would be damaged by the irrational nature of behaviours which must be prevented, even though the individual wants to keep the initiative over them. 'He will be forced to be free',[4] as the famous saying goes, and he will be compelled to side with reason, even if he needs help to do so, against his expressed will. The carer is called upon, and allows himself, to support the part with which the patient does not currently identify, but would identify with if he really and correctly supported his role as a person. It is thus possible to deprive of any responsibility a person who threatens the existence of a company or the fortune of a family.

In these three cases, which have only been sketched and of course are in no way exhaustive, the role played by the concept of the *person* is obvious. Under the pretext of declaring and maintaining values which it claims to originate within itself, the person covers with its mask some eminently social values that absolutely have not originated in the individual, and to which he must submit himself. By the distention which is inherent in it, it allows society ostensibly to defend *his* values rather than its own in relation to the individual thus transformed into their scene, but in fact to impose *its* laws, by paying no heed to the consent of individuals who actually cannot give it, and to defend those values without any nuance, without any qualification, and without any delay.

We are certainly not denying those risks, but we should take the trouble to investigate whether there are not better ways of dealing with them than accepting

[4]Rousseau's declaration was made in another context, that of *The Social Contract*.

the humiliating conditions or uniformization of the 'person'. This is the point where we encounter probabilities and the specific management of situations involving them, if only we can overcome our principled refusal not to take them into account.

2.2 The Attitude Towards Risk as a Mode of Management of Probabilities

The management of probabilities is never simply a theoretical problem. It is always at the same time a practical problem, since it is precisely when one does not know with certainty how a situation is going to evolve that the question of the probability of that evolution arises, especially when the situation demands a decision to act, or to refrain from acting, the outcome of which decision will modify it anyway. That practical issue is never without ethical consequences.

In any field of medicine, the carer has access to statistics and can, thanks to them, define an attitude about any patient in a given situation, based on a history the phases of which he more or less knows. In any case, the indications provided by statistics are important, but in order to be correctly treated, they require a tailored adjustment to a particular situation, because what happens for an individual is never what happens for the fiction of the average individual of statistics. They are delicate and difficult to handle ethically, for the patient should not be given false hopes by a gauche statement of relevant data, nor should any hope be compromised by letting him understand that his case is among a quartile or decile where things are neither going well nor going to get better. However, the use of statistics to define a prognosis or a treatment may be even more difficult in psychiatry than elsewhere in medicine. Indeed, the individualization of the doctor who makes the diagnosis is even more crucial here than elsewhere. Where the psychiatrist would need, more than the other doctors, trustworthy statistics, since everybody expects him to provide valid predictions, he seems to have fewer means to get them than other doctors. However, it is mainly on him and on his colleagues that the responsibility of the decision of discharging a patient or not admitting him to hospital, and its consequences – if that discharge or that non-admission go wrong – will lie. That decision takes into account multiple aspects, which, despite their heterogeneousness, must be reduced to a sort of common measure on the same scales in order to decide which of those aspects must prevail over the others, or how to reach a decision out of the combination of the different variables. For the patient may have wishes or a will. The family may have its own wishes and will, and it has its own interests to be defended, supposing they are homogeneous. Society – 'big' society – also implicitly requires that its interests, and especially those of its safety, be taken into account. Carers may moreover find advantages and drawbacks in the treatment, and it is the same for the other patients who will be the unwilling partners of the patient whose presence is imposed upon them. These elements, which are from the beginning greatly heterogeneous, must be assigned a value, as if they could be reduced to a sort of common denominator, by taking into account a projection towards the future of each of those elements, which is called a probabilization. From which points of view should the calculation be made? Undoubtedly, if a member of the family had to do it, he might not weigh the elements he takes into account in the same way, and would not get the same result as the patient himself or as the other patients if they had to make a decision, or were given the opportunity to take one. Should these points of view be left to confront one another in a game which must be concluded somehow, since it cannot be left without a solution? Should the selection of the best solution be left to be carried out through some kind of filtering, whatever its principles? Should the pivotal point, if there is one, be chosen at the outset? Moreover, let us not believe that the viewpoints and decision centres, from which the filtering or balance is established, can be defined in advance. They constantly evolve as the situation develops.

In any case, whether the solution is taken to be a filtering or a composition, the complexity of the task is such that the carer is tempted – especially if it is an emergency situation and he does not have much time to take his decision – to fall back on the safest solution, be the safety in question, depending on the situation, that of the patient, of the other patients, of the family or of others broadly speaking. The problem, for us, is to find out whether this way of proceeding is the most strategic one, on the one hand, and the most ethical one on the other. To start dealing with it, and to show the beginning of a solution, we shall rapidly rehearse the previous three cases: namely that of the risk of suicide, that of the danger to others, and that of the risk for all sorts of social issues.

In adopting the 'safe' strategy, if one thinks one has indications, through the speech or behaviour of someone, of *his propensity to commit suicide*, one will be tempted to assume the worst, as if the individual who exhibits such indications were really about to commit suicide, and consequently will try to annihilate, at least for some time, the bare possibility that he acts and thus eliminates any probabilities of that act existing. Is such a way of dealing with the issue rational? Why should the degree of probability of an act be replaced by the certainty that it will or will not happen, while that degree is precisely uncertain? It will be objected that, in the case of suicide, there are only two possible positions: either it is attempted (even if it fails) or it is not. Yes, but why systematically assume the worse solution, according to which it will be committed, and in response detain the potential candidate for suicide in hospital, remove his shoelaces, and sometimes all his clothes etc.? This option is much less rational than may seem. It presupposes that the doctor knows how the situation is going to evolve better than the patient, that he does not trust him (in the case where the patient has given his word), and that he has the right or even the duty not to trust him, as if the patient were systematically and completely disqualified in his role as a partner to a contract, and as if his words had no grip on reality, neither present nor future. No heed is paid to Nietzsche's words,[5] as if protection through prohibition was infinitely more important than the assessment of his own life by the

---

[5]This allusion is detailed in note 10 below [15–17].

individual himself, who is once again disqualified in that assessment which, after all, concerns him so closely. One stumbles over that famous protection of people against themselves. Why would the right to kill oneself, that most of us do not question for other men, be absolutely and out of principle prevented and denied to someone whose judgment and will are thought to be so impaired that his life – which is his, nonetheless, and which he thinks is degraded – has to be protected against him? Why should the reaction to that presumed impairment always be in the direction opposite to that which the patient wants?

Our *second* case is more delicate, it is true. It seems to be rationally well founded to assume the worst when others are to be protected against a potential assault. That protection would be a duty, on the pretext that those others have not asked to be exposed to additional risks and have not consented to be so, since, most of the time, they are not even aware of them. The doctor can therefore reasonably think that he does not have to right deliberately to expose others to a risk of being assaulted vastly in excess of the risk they would run if the patient remained detained. Here again, however, it is much less rational than it seems to be, as if – in a distribution of possible situations or cases of possible evolutions of that situation – it were always the worst solution or case that should be chosen as the one take as a reference to deal with the situation in its entirety, without gauging the degree of probability, and proceeding as if that gauging were no longer relevant, as if it should not be taken into consideration when set against the possibility that the patient would claim victims – a possibility which is not treated as a possibility, but as a fact the probability of which is almost 1 (close to certainty). Is that bias strategic? Is it ethical? In other words, is it rational, in the sense of something which is pragmatically defendable, and is it *righteous* or *equitable*?

The *third* case is that of a social good, whether economic, cultural or educational, which may be compromised if not destroyed by ill-considered decisions. Curiously enough, this case may be the one on which everyone can agree most easily, because it makes the well-being of people depend on decisions which may be abnormal and which certainly put their living conditions – if not their very lives, health or tranquility – at risk. One may accept the need to defer to external authority in political, administrative or economic life, but it is expected to be rational, even if that rationality includes different degrees and interpretations. It seems that, on this point at least, the probabilistic and utilitarian ethics that we are about to defend is in harmony with the 'personalistic' perspective that we are criticizing.

Our criticism of the 'personalistic' perspective now has to go much further in order for us to understand, at first hand, in that field, that the refusal to take probabilities into account, in a differentiated and graduated way, in order to solve problems that involve them, is not only theoretically wrong, since it does no more than reverse the prejudices of those who assume certainties without sufficient warrant or act as if they did so, but also because its bias is unfair, practically speaking.

2.3 Furthering the Criticism of the Personalistic Approach

In the scientific field, Descartes recommended a doubt that went so far as holding as false all that was only probable. In other words, as Leibniz could legitimately reproach him,[6] on the ground that instead of calculating the degree of probability of some statement and treating its truth as some fraction of certainty, he preferred rejecting it as false, believing that he would thereby avoid being misled into holding as true that which was dubious in any degree, and therefore presented the risk of being false. Strange and almost absurd as that method was at the theoretical level, what was it worth at the practical level? Descartes did not recommend it at that practical level. He did not ignore the way in which such an attitude would lead to scepticism, hesitation, and, in the end, to widespread inaction, which would consist in letting the others act in one's stead, without one doing anything oneself. Resolution in one's action is a indisputable virtue.[7] However, the author of *Discourse on Method* then advised his readers to stay away from extreme certainties in one way or the other,[8] and to limit themselves to actions well received by our institutions and the general public. Descartes' advice is not convincing, however, for, once again, as in the theoretical domain, he paid no heed to the assessment of the probable and of the degree of probability of the elements of a situation, and in the name of prudence only looked for a sort of in-between which ignored them; and, which if adopted, would only lead to conformism and under-considered, inappropriate, actions, for they

[6]In his *Reflection of the general part of the Principles of Descartes* [13], Leibniz remarks on Article 2, which poses as a criterion for a good rational method that 'qu'il est utile de considérer comme fausses toutes les choses dont on peut douter' < it will be useful to consider as false what is doubtful>: 'Je ne vois pas à quoi sert de tenir les choses douteuses pour fausses. Ce ne serait pas là se dégager des préjugés, mais en changer. Que. si l'on ne veut y voir que fiction, il ne fallait pas en abuser' ([12], p. 288). <I do not see of what use it is to consider doubtful things as false. This would not be to cast aside prejudices, but to change them. But if fiction is so understood, it must not be abused> [11]. In other words, there is an ethics of fiction and it should not be introduced haphazardly.

[7]The second maxim of the Third part of *Discourse on Method* which defines 'quelques règles de la morale tirées de la méthode' is 'd'être le plus ferme et le plus résolu en mes actions que je pourrais et de ne suivre pas moins constamment les opinions les plus douteuses, lorsque je m'y serais une fois déterminé, que si elles eussent été très assurées' < to be as firm and resolute in my actions as I was able and not to adhere less steadfastly to the most doubtful opinions, when once adopted than if they had been highly certain> [6].

[8]'Entre plusieurs opinions également reçues, je ne choisissais que les plus modérées, tant à cause que ce sont toujours les plus commodes pour la pratique, et vraisemblablement les meilleurs, tous excès ayant coutume d'être mauvais; comme aussi afin de me détourner moins du vrai chemin, en cas que je faillisse, que si, ayant choisi l'un des extrêmes, c'eût été l'autre qu'il eût fallu suivre' < Amid many opinions held in equal repute, I chose always the most moderate, as much for the reason that they are always the most convenient for practice, and probably the best (for all excess is generally vicious), as that, in the event of my falling in error, I might be at less distance from the truth than if, having chosen one of the extremes, it should turn out to be the other which I ought to have adopted> (idem). Which is what is recommended by Rawls in the principle of the maximin.

would have no chance of taking cognizance of all the conditions of the situation. This advice is equivalent to allotting any option whatsoever, so long as it is not extreme, the probability of ½, and to allow no change in that probability, even though the evolution of the situation or of our knowledge of it meant that its probability became only ¼ or $^9/_{10}$. It is true that Descartes did not himself endorse a type of practice that we have seen proliferate in the last decades of the twentieth century – especially on issues of social justice – and which consists in applying the attitude of blanket disregard of probabilities that he had adopted in the theoretical domain. It consists, as we have already started to show and criticize, in remaining insensitive to the degrees of probability and, behind a veil of ignorance, for we do not know the developments of the situation, to assume the worst in a situation by systematically siding with those who, in a given situation, would suffer most, and would derive least advantage from what we are proposing to do. If what we are about to do is acceptable to those who would gain least from it, then the action is acceptable. If what we are about to do involves an unacceptable risk for those 'victims' who would be most severely affected by it, then that action should be rejected, even though it might be better in other ways than the alternative actions. In other words, what claims to be *rationality*, in these cases, consists in doggedly adopting the point of view of the victim, without taking into account the parameters of the probabilities of the consequences of an action, and without any other explanation or rationale.

The absurdity and false prudence of such an attitude towards the probable arise from two sources. *First*, they are entailed by the determination to ignore degrees of that probability, and to treat it as if it were a complete certainty or – and this amounts to the same thing – as a radical uncertainty. In other words, one makes a mistake deliberately and consciously, and with an obstinacy that claims to be virtuous. It is quite possible, however, in relation to an action by which one wants to reach a goal, given the statistics, or past experience, that one grants some parameters a provisional degree of probability and rectifies it along the way, in case it has been over or underestimated. This flexibility and intelligence in action seem superior to us to allegedly prudent approach which has led to the excesses of the precautionary principle.[9] That is exactly what is called *running a risk*. There is however a *second* way of unmasking the absurdity of that false rationality. Systematically siding with the potential victim in a situation – adopting the viewpoint of him who would suffer the most from the situation – consists in arbitrarily reducing the action to only some of the aspects or fragments which constitute it, and perceiving only some of its aspects, while obstinately refusing to see all the others, rather than taking it holistically, in its entirety. How can one say they are on the side of reason when they refuse out of principle to take into account all the elements involved in a situation, at least given what one can know of it?

---

[9]Bentham had already seen in what chain of events that principle crushed us: 'In proportion as the rule is safe, secure against being productive of erroneous decision, it is in the same proportion useless. Safe, it is not effective; effective, it is not safe' ([2], 226).

It might be argued that it is precisely because one does not know how situations develop that one should be prudent and make as few victims as possible, especially when the lives or deaths of people are at stake, and introduce as little evil into the world as possible: this would be to misunderstood our point. There is no evidentiary basis for assuming that, by systematically choosing the viewpoint of the potential victim or victims in a situation, one will choose the point of view that is best adapted to the situation, which is likely to bring more good to the world by taking the most reasonable risks. We are talking of probability precisely because we do not know the whole situation, and above all we do not know how all its developments will play out. Nevertheless, even if we ignore the situation, because it is not a case of complete certainty, it is not always a case of a complete uncertainty either. There are potential developments of variable plausibility which may allow us to be more or less confident of intermediate points, and which excuse us from confining ourselves to concentrating on one possible point of view and sector of the situation as if they were the whole of it. There is a false rationality of prudence, or, more precisely, there exists an appearance of rationality of a false prudence, by which one should not be fooled.

Many other aspects could be put forward in criticism of the 'timorous' rather than prudent, the partial rather than rational, rationality, and the 'personalism' that very often lies behind its dogmatism. The attitude that we have just criticized favours some decision centres to the detriment of almost all the others, and it stabilizes them in some roles to the exclusion of all the others, in a dogmatism that is shrouded in prudence. Why should only a few individuals – or even a majority – have the right to be treated as persons, here and now, while others – without any reason being given – are to be denied that right in the same present, but only by reference to a future where they will supposedly be happy that we have taken the decision in their stead? Here again appears the selective, more or less clandestine, character of an attribution of the concept of person which cannot be explained by the choice of who should benefit from that status and who will not benefit from it, or will do so only in a 2nd, 3rd or nth degree, should the opportunity arise. Why should the choice of preserving the potential victim be more conformable with the concept of person and with reason?

Admittedly, for a finer, more delicate and less dogmatic rationality to operate, laws should be less damning for the doctor who has taken a decision which has revealed itself to be unfortunate in that it has opened the possibility for the patient to commit suicide, or it has rendered the lives of others impossible, or even because it has allowed them to be assaulted, or become poorer, or be deprived of well-being. One should understand then that, in his situation, based on the information he had, and given the interests of everybody in a given situation, he did not necessarily take an imprudent decision, even though its consequences have been unfortunate. It is not for laws to encourage audacity. They are even made to dissuade it, but precisely not to the point of annihilating any risk and globally producing more displeasure than pleasure. If an individual commits suicide after a decision of his doctor, who, it is discovered with hindsight, has trusted him too much, is the mistake systematically the doctor's? Undertaking risky endeavours does not mean they will succeed. Assuming a risk does not take away the possibilities of a failure. Probability does not become certainty from the moment it is assumed. Is it ethical to transform what has ended up badly, but could have ended up happily, into a fault of the doctor? Is there not more wisdom in Nietzsche's words, when he recommended neither pushing into the act whoever is thinking of committing suicide, but nor seeking to prevent the act in all circumstances, thereby abolishing any right to kill oneself?[10] If an individual assaults another, after the decision of the doctor who trusted his patient's words or behaviour too much, should the doctor out of principle be held liable when he tried to find a balanced position, taking into account all the elements, the development of which seems to have demonstrated he was wrong, even though he might not have taken the wrong decision, given the information he had? Is it not as serious a mistake to make somebody suffer uselessly, who may not make other people suffer, in the name of preventive precaution? The costs of this latter mistake are invisible however. They cannot be established, since the idea of dangerousness prevents them from being seen, and their value has been effectively reduced to nothing in advance.

It is quite clear that we are substituting to an ethics of the person an ethics of calculation and utilitarianism. The *first* does not allow calculations and cannot endure giving reasons to favour one person – for everybody may claim the status of *person*, and it is granted or taken away by ethics itself, which recommends it only for reasons which have nothing to do with the very idea of *person*. In addition, by a principle partiality, it disqualifies us from looking at the whole situation purely because of our reverence for the idea of the person. The *second* is less interested in persons than in the mobile and very variable centres of pleasure and pain – of preferences as contemporary utilitarians would say – which evolve as the situation evolves, the aim being not to favour one person or another, but to look at how to arrange what seems to give the most possible pleasure to the partners as a whole who are being united in and by the situation, by taking into account *all* the aspects of the situation, given the probabilities of evolution of each of those elements.

We do not think we have solved all the difficulties with this displacement. Yet at least we no longer fall into the trap of a personalism, which creates misery more than

---

[10]'THE PREVENTION OF SUICIDE. – There is a certain right by which we may deprive a man of life, but none by which 29 we may deprive him of death; it is a mere cruelty. [. . .] OLD AGE AND DEATH. – Apart from the commands of religion, the question may well be asked, Why is it more worthy for an old man who feels his power decline, to await his slow exhaustion and extinction than with full consciousness to set a limit to his life?' ([15], §§ 88, 80). A few years later, in [17], § 36 of 'Skirmishes of an Untimely Man', Nietzsche wrote, in a statement that was at the limit of the unacceptable transgression, in 'MORALITY FOR PHYSICIANS': '(. . .) In a certain state it is indecent to live longer. To go on vegetating in cowardly dependence on physicians and machinations, after the meaning of life, the right to life, has been lost, that ought to prompt a profound contempt in society. The physicians, in turn, would have to be the mediators of this contempt – not prescriptions, but every day a new dose of nausea with their patients. To create a new responsibility, that of the physician, for all cases in which the highest interest of life, of ascending life, demands the most inconsiderate pushing down and aside of degenerating life – for example (. . .) for the right to live. To die proudly when it is no longer possible to live proudly (. . .)' ([16], p. 151). http://www.inp.uw.edu.pl/mdsie/Political_Thought/twilight-of-the-idols-friedriech-nietzsche.pdf.

it makes people happy. On the second view the idea of *risk*, which implies a behaviour adapted to the degree of probability or hope we rightly attach to the development of the situation, the idea of *utility* and the game of pain and pleasure that it implies, and that of the *centres* to which those pleasures and pains apply, seem in tune with one another, and offer a good enough grip on reality to be perfectly capable of competing with personalism. Does that mean that any concern for the person should be radically eliminated? Probably not, but we still need to try and rein it in.

Let us pause, before doing so, and consider two features. First, Risk is not adventure, it is the opposite. The adventurer – who does not have any choice if he is in a situation where he is forced to be one – does not know where he is going and moves forward in quite uncertain circumstances, or at least he treats them as radically uncertain and as not allowing any prior assessment, even though he could find himself in situations in which he could make such an assessment and transform his adventure into a situation of risk. For whoever takes risks does not evolve in a world in which he knows everything with absolute certainty, but puts projects forward combined with degrees of certainty which may be ascertained, and which he could account for if things went wrong, which, is – unfortunately – always possible. One can, quite prudently, take risks, whereas adventure is always imprudent. Prudence does not consist in not taking risks, but in taking them – even if they are really major – in a measured fashion and in such a way that the measure will be acceptable. There are cases in which we have no choice but that of adventure, and in which adventure is obvious for us. Fortunately, they are quite rare in medicine – they may exist in a case of extreme emergency – but should not be taken as the general case. On the other hand, cases in which people take risks are quite frequent, and even constant. It is important not to dispose of them quickly as simple uncertainties, under the pretext that they are indeed not certain. Rationality lies in the search for degrees of certainty – be they quite objective (as when it is possible to establish statistics and trends), or subjective (in cases where it is not possible to get such objectivity and a change of viewpoint is needed, which does not hinder the possibility of calculating our actions) – and in our actions' conforming with our calculations.

Second, mathematicians draw a distinction between *subjective* or *Bayesian probabilities*, which attach to the project one considers in a given situation, and *objective probabilities*, which seem to attach to the elements of the situation. The risk is of course rather on the side of subjective probabilities, but it can be assigned some sort of objectivity to the extent that it leads us to apply a coefficient to the elements of the situation in which it develops. The coincidence – which should not be taken for an identity – between the concepts of risk and of probability arises from the fact that probability is different from the objects it seems to designate and quantify. Despite appearances, probability does not mention the die or the sides of the die. Its object is not, properly speaking, in experience. It does not coincide with anything that can be found in experience, or at least not in a unique experience coincident with reality.

Pascal put it perfectly.[11] But the object it is about and which is behind the objects of experience is such as one can grasp through action: it can form part of the aim of an action. Probability is a rhetorical concept about action. Risk is the other name used to describe the way in which action grasps the object of probabilities.

2.4 Substitution of the Concepts of *Trace* and *Individual* to That of *Person*

It is not for me to say whether the concept of *person*, and the defects we have found it has, may have some therapeutic value in psychiatry. That point should be considered in relation to each illness, and, in this sector of medicine, in relation to each patient, since individualization is so extreme in this field. We are ready to grant ethical properties, which are not always found in the centres of pleasure and pain that we are trying to substitute to it, to this concept of *person*. It is easy to put it into time as it lets itself be distended by time factors. It is not inessential to point, through the person, to the possibility the individual has in himself to invent himself, more easily and possibly to better effect than the ideas of pleasure and pain would allow. Lastly, it allows a distension between *what* the individual is, has been and will be, and what he thinks he *must be*. These characteristics are not insignificant, and are not all transposable to the set of ideas that we suggest as a replacement for the concept of *person*, which is why it is necessary to sketch it, in any ethical situation, even where it leads only to small sections of reality, without any overall perspective.

This is all the more the case as neither the isolation it represents in a situation, nor the unilateral nature of the cutting it produces there, should be exaggerated. The specificity of the *person* is to try and adopt the point of view of the other, and only take a decision when that point of view has been assumed. It is also this difficulty of perspective which favours the play of pleasures and pains. For even though the happening of pleasures and pains in a situation, be they actual or contemplated, could be attributed to different centres, they can be added – and can lead to calculations – only if they too are related to a common denominator, which is not that easy to find. Admittedly, the question is not to draw a distinction between real and false pleasures, as personalists make a difference between the 'true self' and 'false self'. Pleasure must be taken into account from the moment it acknowledges itself and values itself as such. It should not be neglected under the pretext that it is false. In any case, no one can set himself up as a judge of the pleasures and pains of the other, unless his own pleasures and pains are at stake. The problem of their

[11]Indeed, in his address *Celeberrimae Matheseos Academiae Parisiensi* (1654), Pascal indicates that in order to *participate in games*, 'il faut chercher [à résoudre le problème] d'autant plus rigoureusement par la raison que les possibilités sont moindres d'être renseigné par l'expérience. En effet, les résultats ambigus du sort sont à juste titre attribués plutôt au hasard de la contingence qu'à une nécessité de nature' ([18], p. 1034–1035). This text is quoted more extensively at the end of this book.

commensurability is no less delicate: how are they to be assigned their respective weights? Which unit should be chosen to perform that sort of weighing? Who should do it, and from what point of view?

The difference in resorting to intersubjectivity or to universality is however quite great from one perspective to the other. Intersubjectivity, when it is performed within the framework of 'personalism', is obtained through the consideration of whether humanity in its invention benefits from an action or a decision, whereas the point of view of pleasure and pain only requires that an individual put himself in the place of another or others, to try and understand how it is ordered – without judging it – or to implement it among other pleasures and pains. In a way, and quite paradoxically, the utilitarian point of view seems closer to the Jewish and Christian saying *Do not do unto others what you would not have them do unto you* than personalism, which has, however, often claimed to adhere to Christianity,[12] but which in fact is quite far from it, especially when it takes the form, as in Kant, of its projection onto a sort of consistent universe, a 'nature', understood to be an intertwining of ideal laws, the possibility of the unity of which is cheaply obtained. The consistency of utilitarianism is achieved by a game of balances which must be calculated, and not by the possibility of constituting an ideal universe deprived of probabilities.[13] Utilitarianism is so much linked to probabilism that it cannot endure for ethics to be measured by a sort of 'cannon' like that which is offered in Kantianism in the idea of 'nature', simply because such a 'nature' cannot exist in the eyes of those who have a conception of the law which is only probable, that is, without ideal perfection.

There is another difference upon which we would like to finish before concluding this chapter. Pleasures and pains are not referred to persons, in the works of utilitarians, but rather to individuals or groups of individuals. This difference will be found quite small, since according to a possible view, there is not much difference between a person and an individual and we could talk about them indifferently. The individual however, who refers to his person only refers to it as an abstract idea of himself. Pleasure and pain point to the individual who enters upon the stage as a vivid and singular trace, which is what should be taken into account in ethics. Despite appearances, and the example of Mill is the best testimony of it,[14] ethics is

[12]Starting with Gabriel Marcel, the founder of 'personalism' himself, who understood it in a more specific meaning than that which we have given it up to now.

[13]The promotion of the maxim of his action to the status of law does not imply any calculation nor the taking into account of the degrees of probability.

[14]In a letter to E. Acollas of 20th September 1871, Mill recalls that the principle of his *On Liberty* is that of the 'autonomy of the individual'. Indeed, the fifteen pages of Chapter III of *On Liberty* are about 'Individuality, as One of the Elements of Well-Being', and the last five pages of the conclusion are about freedom as individuality. 'Je reconnais', he says in French, 'cette autonomie comme une règle rigoureuse dans les choses qui ne regardent que l'individu lui-même ou, si elles intéressent les autres, qui ne les intéressent que par l'influence de l'exemple ou par l'intérêt direct que d'autres peuvent avoir au bonheur et à la prospérité de chacun. Par cette doctrine, j'affranchis de tout contrôle, hors celui de la critique, le cercle de la vie individuelle proprement dire' ([14], pp. 1831–1832). Those sentences could never have written by Kant for whom a person gives the individual duties and does not free him from any control without failure. In another letter, from the

not the same when the self is taken to be 'personal', as in Kant, and when that self is an individual self who feels pleasures and pains, who is anchored in being through their play itself. An individual may be either a sequence or a series of those traces pointed out by pleasures and pains. The idea of trace gives the individual more existence than that of person, which does not care any more about the idea of the individual than it cares about the ideas of pleasures and pains. The idea of trace is to be found in the work of Jankélévitch [9].[15] The only delicate point in the idea of the trace of writing or sign that the existence of a being is, is that the latter may 'exist' too much to have any ethical importance. It 'exists' in any case more than it 'must exist', and that may be where the person retains an advantage, which we are not trying to contest. Its being torn between be and must-be – the fact that it is a sign anchored in being on the one hand, and a perspective of must be, having to invent itself, on the other – may be an advantage over a philosophy that relies on pleasure and pain as if they simply « were », without permitting enough space for the perspective of « must-be », which seems difficult to do without in ethics. The fact of existing, which offers a sort of particular ontological resistance, is not a great help in shaping the void which is that of our existence, and, perhaps, through its lesser 'existence', the person manages to do it better. In *Ethics*, G. Moore showed that an ethics of pleasure and pain does not necessarily provide any duty, for there are several ways of solving the balances between pleasures and pains in a situation, and there are undoubtedly more ways than one of getting a maximum of pleasure in a given situation. The weak point of utilitarian ethics may be that they have only a very lose idea of duty, especially when it is understood as that which conforms to the principle of utility – that is, as that which prescribes the maximum of happiness for the greatest number[16] – and not as a tension between what we are and what is before us, which is still non-existent and yet to be created.

---

same time, of 20th January 1871, addressed to the New York Liberal Club, Mill supports the value of the individuality of thought and character, combining it with the possibility that it be criticized, 'There cannot be a higher or more important air than of assertion & maintaining individuality of thought & character, together with its necessary complement, the fullest latitude of mutual criticism' (ibid., pp. 1801–1802).

[15]What exists of what is called a *person*, which properly carves the being and maybe gives it some sort of eternity, is the 'fact' that it *has been*, since even when it is no more, the event that it has been is impossible to take away from it, without it being quite possible to know from whom it is impossible to take it away. Some resistance, some necessity, some hardness thus happens through that frailty and that *having been* does not only work upon those who, as the saying goes, remember the dead, but – even though there were nobody to remember them – upon existence itself which it structures and to which it gives specific duties. (See [4]). Our first volume, published last year, addressed this point: see *Rethinking Medical Ethics*, ([5], pp. 35–61; pp. 139–160).

[16]See Bentham J.: 'Of an action that is conformable to the principle of utility, one may always say either that it is one that ought to be done, or at least that it is not one that ought to be done. One may say also, that it is right it should be done; at least that it is not wrong it should be done: that it is a right action; at least that it is not a wrong action. When thus interpreted, the words *ought*, and *right* and *wrong*, and others of that stamp, have a meaning: the otherwise, they have none.' ([1], p. 2, §X).

That is why, despite all the weaknesses of the concept of *person*, we do not think that it would be reasonable to try and get rid of it completely in ethics. One thing in favour of theories of pleasure and pain, which are the only ones to give probabilities some meaning – whereas the ethics of the person cannot and remains paralyzed before the concept of *risk* – is that by using them it is possible to reintroduce the concept of *person* one fragment at a time. This may have been what Mill tried to do in relation to Kant. What is certain is that it is easier to introduce Kantianism requirements into a utilitarian perspective than to introduce utilitarian perspectives into Kantianism, which radically refuses them, even if it unwillingly yields to them *in extremis*. That refusal is not found on the opposing side, for, as Harsanyi showed, if you want to live according to Kant's precepts, and if that is your pleasure, provided you do not want to impose that pleasure upon others as the only one that suits them, it is always possible in a utilitarian context. There is a place in utilitarianism for those who find pleasure in making themselves suffer. There is little or none in Kantianism for those who prefer their pleasure to the categorical imperative, even though that pleasure were no inconvenience for the other.[17]

2.5 Conclusions

We would like to finish this chapter on the reversal of perspective we have announced, by recalling that up to now we have constantly moved forward, as if psychiatry had a separate place in the ethics of care from all the other sectors of medicine. In particular, we have seen that the principle of autonomy has been challenged in a variety of ways in psychiatry. It may be a mistake though, to believe that that principle can very easily, in actuality and without rhetorical sleight of hand, accompany the other sectors of medicine. In reality, psychiatry highlights the impossibilities and contradictions of the concept more blatantly, while everywhere else those impossibilities and contradictions are masked or well disguised. What we get when we think about ethics in psychiatry can contribute to the criticism of the principle of autonomy even where it seems essential to the point where the law itself has given it a central place in the medical system.

That appearance becomes fragile as soon as one is faced with a serious illness. Admittedly, the patient may have his care terminated if he wants, and, provided he is properly informed by his doctor, the latter's responsibility stops there. Even though it is demonstrating respect for autonomy in that case, everywhere else, that is, in the majority of cases, it is difficult to talk of the autonomy and responsibility of the patient. Indeed, *either* he is placed before a choice between false alternatives – he

---

[17]At the same time, it could be possible to show, as Harsanyi has done, that Rawls' principle of maximin is but a particular case of the Bayesian approach to ethical problems that one should have. From time to time, the Rawlsian approach is in harmony with the Bayesian. The latter can explain why it is so, but the reverse is not possible. When the rule of the maximin is followed in isolation, it is 'a highly irrational decision making rule'. see Harsanyi, ([8], p. 47).

knows the advice and preference of his doctor, and, if he is rational and at the same time rejects the option of trying to recover his health on his own, he will eventually follow his doctor's advice which, in the end, will seem to support the best solution – or the alternative is a true one, in which case it is difficult to see how the patient, if he is not competent in medicine, which is most often the case, could decide what doctors themselves have difficulties agreeing upon, and how that patient's decision could have any value in the eyes of the doctors or indeed of anybody else. The patient's autonomy in relation to decisions about his treatment can really only be in *trompe l'oeil* style, and one cannot see how it could be any different. Psychiatry shines, at least in certain cases, a stark spotlight on this point.

Moreover, according to a bias in favour of *autonomy* and to the concept of *person*, we are supposed to receive from doctors the best treatment, that which would respect our humanity the best. Though that is often the case in political and educational issues,[18] I am not sure it is the case in illnesses. It is possible, in the name of what one owes oneself when one is a person, to have to face terrible news which may be announced without any precautions, provided that pleasure and pain are less essential values than the taking into account of the person and the concepts that accompany it. Ethics, as conceived on the basis of those concepts, may inflict a form of violence against oneself, or against others, with the best conscience in the world. The concept of *person* does not necessarily do any good to the individual who is made prey to its demands. It may remind him of the requirement to conform to the *right* without doing him any *good*, in circumstances in which the requirement of *right* is in no way necessary. Is that violence necessary? If violence is often lamented – because autonomy and person, consent, whether enlightened or not, do not always have any role in it – is it not possible to find some violence at the opposite pole, when we are bluntly informed of our prognosis and reminded of our duty with sadistic relish – though it may be the doctor's legal obligation to inform us – or when we inflict it upon ourselves with morose masochism? This shows, in passing, that violence does not necessarily address the person, nor derive its meaning only with reference to the person.

I would like to end with three paradoxes which we have constantly met during the investigation in this chapter. The *first* is that there is a certain *Don Quixotism* of the person that we cannot tire of observing. Those who defend the concept constantly side with it as if it were the reality of which the individual who plays his desires against it must be constantly reminded. It defends itself only through incredible chimeras which make it lose any contact with reality, and consequently any credibility. One ends up not knowing whether it is the illness or the supposed remedy that is the worst madness. The *second*, more 'positive' paradox, is the confirmation that utilitarianism, which could not align itself with Jankélévitch anymore than he could

[18]This first assessment, which is being roughly sketched without refinement, requires further elaboration and lacks proper nuance, which might be supplied by further reflection on, for instance, the idea of *merit*, which lies at the heart of educational values, and involves extremely complicated and contradictory relations with the concept of *person*, which most often does nothing more than muddling it.

align himself with it, is nonetheless much closer to the doctrine of those, like him, who defended the idea of the indelible *trace* of the individual in ethics, than is a personalism to which one could have thought, before reading Jankélévitch, that he was closest in his way of considering ethics. *Thirdly* and lastly, and in a way related to the previous paradox, one might be surprised at seeing a doctrine of calculation so close to a humanism of the individual and of trace. Calculation, as Plato pointed out long ago, is deeply human, maybe even more archaically so than speech and writing,[19] and the defence of the trace necessarily uses calculation, as Jankélévitch showed [9].

Our approach of probabilities and of their calculation in ethical questions leaves before us an issue that will be studied in the next chapter. Potentiality is criticized in philosophy, especially by analytical minds which ordinarily show its inconsistency and contradictions. However, does what we have said about probabilities not give an acceptable meaning to a concept which is admittedly difficult to do without in ethical questions?

Bibliography

1. Bentham, J. (1843). An introduction to the principles of morals and legislation. In T. Bowring (Ed.), *The works of Jeremy Bentham* (Vol. I). Edinburg: Tait; Bentham, J. (1970). *An introduction to the principles of morals and legislation*. J. H. Burns & H. L. A. Hart (Eds.). London: Athlone Press.
2. Bentham, J. (1843). Pannomial fragments. In T. Bowring (Ed.), *The works of Jeremy Bentham* (Vol. III, p. 226). Edinburg.
3. Bouveresse, J. (1993). La théorie des fictions chez Bentham. In K. Mulligan & R. Roth (Eds.), *Regards sur Bentham et l'Utilitarisme*. Genève: Droz.
4. Cléro, J. P. (2017, Juillet). Une pensée de l'existence à l'épreuve de l'éthique des soins. Les contradictions de l'éthique médicale. In *Jankélévitch; morale et politique, Cités*. Paris: PUF
5. Cléro, J. P. (2018). *Rethinking medical ethics*. Stuttgart: Ibidem-Verlag.
6. Descartes, R. (2018). *Discourse on method* (1637), Trails by J. Veitch. http://pinkmonkey.com/dl/library1/book0648.pdf
7. Harsanyi, J. C. (1980). *Essays on ethics, social behavior, and scientific explanation*. Dordrecht/Boston: Reidel.
8. Harsanyi, J. C. (1982). Morality and the theory of the rational behaviour. In A. Sen & B. Williams (Eds.), *Utilitarianism and beyond*. Cambridge/Paris: Cambridge University Press/Éditions de la maison des sciences de l'homme.
9. Jankélévitch, V. (1983). *L'irréversible et la nostalgie*. Paris: Flammarion.
10. Kant, I. (1996). *The metaphysics of morals, part I. Metaphysical first principles of the doctrine of right*, Intr. III, in Kant I., *Practical philosophy*. Cambridge/New York/Melbourne: Cambridge University.

---

[19]Plato, *Epinomis*, 977c: 'if we should deprive human nature of number we should never attain to any understanding. [. . .] and the creature that did not know two and three, or odd or even, and was completely ignorant of number, could never clearly *tell* of things about which it had acquired sensations and memories' ([20], p. 439).

11. Leibniz, G. W. (1890). 'Animadversiones on Descartes Principles of Philosophy', books 1 and 2. In *The philosophical Works of Leibnitz, comprising the monadology, new system of nature, principles of nature and grace, letters to Clarke, refutation of Spinoza and his other important opuscules, together with the abridgment of the theodicy, and extracts from the new essays on human understanding*. Tuttle: Morehouse & Taylor. http://www.archive.org/steam/philosophicalwor00leibuoft_djvu.text
12. Leibniz, G. W. (1972). In L. Prenant (Ed.), *Oeuvres* (p. 288). Paris: Aubier Montaigne.
13. Leibniz, G. W. (1989). *Critical thoughts on the general part of the principles of descartes* (1692). link.springer.com>content>pdf. Reflection of the general part of the principles of descartes.
14. Mill, J.-S. (1972). *Collected works of John Stuart Mill*, vol. XVII, The Later Letters of John Stuart Mill, (1849–1873), London, Toronto, Routledge & Kegan Paul, 1972 (1849–1873). London/Toronto: Routledge & Kegan Paul.
15. Nietzsche, F. (1910). *Human, all-too-Human*, §§ 88, 80 (trans: Zimmern, H., & Foulis, T. N.). Edinburgh/London.
16. Nietzsche, F. (1985). *Le Crépuscule des Idoles, suivi de: Le cas Wagner* (p. 151). Paris: G.F. Flammarion.
17. Nietzsche, F. (1997). Hackett publishing company, Indianapolis (USA); *Twilight of the Idols*, http://www.inp.uw.edu.pl/mdsie/Political_Thought/twilight-of-the-idols-friedriech-nietzsche.pdf
18. Pascal, B. (1970). *Celeberrimae Matheseos Academiae Parisiensi* (1654). In *Oeuvres complètes*, Paris: Bibliothèque européenne Desclées de Brouwer, Vol. II.
19. Pascal, B. (1995). *Pensées* (Trad: Krailsheimer, A. J). London: Penguin Books.
20. Plato. (1986). Epinomis. In *Plato in twelve volumes, XII* (trans: Lamb, W. R. M). Cambridge/London: Harvard University Press/W. Heinemann LTD.

Chapter 3
Ethics and the Notion of Potentiality

Abstract The category of *potentiality* is unloved in philosophy because it always seems to imply an implicit and dangerous return to the Aristotelian doctrine which was disposed of in the XVIIth century by thinkers like Galileo and Descartes, whether the issues were theoretical or practical. Yet in a great number of ethical cases, it is not possible to dispense with this category of potentiality that enables us to assess a situation. In thinking abortion, it is impossible to bypass this notion. Is it a means by which a rational account can be taken of *potentiality*? We are facing here the philosophical positions of Singer on a chief subject of his researches; and we will show that his dealing of the notion of *potentiality* tends to dissimulate the dynamics of the phenomena he studies from an ethical point of view. We will try to see the extent to which the notion of *probability* may be substituted to *potentiality*.

Keywords Abortion · Bayesian probabilities · Diodorus Cronus · Hare R.M. · Potentiality · Probability · Singer P. · Vuillemin J.

> 'Mais vraiment, n'ont-ils pas l'air de prononcer des paroles magiques, chargées d'une force occulte et dépassant la portée de l'esprit humain, ceux qui disent que le mouvement, cette chose que chacun connaît parfaitement, est l'acte d'un être en puissance, en tant qu'il est en puissance? Qui donc comprend ces mots?' Descartes, ([3], 2010, pp. 153–154).[1]
>
> 'And, now, the moment. A moment such as this is unique. To be sure, it is short and temporal, as the moment is; it is passing, as the moment is, past, as the moment is in the next moment, and yet it is decisive, and yet it is filled with the eternal. A moment such as this mut have a special name. Let us call it: *the fullness of time*'. Kierkegaard S., ([7], p. 18).

Even though critical minds have cautioned it against the concept of *potentiality*, ethics cannot but make use of it. One of the questions which makes the meeting and

[1] «Again, when people say that motion, something perfectly familiar to everyone, is 'the actuality of a potential being, in so far as it is potential', do they not give the impression of uttering magic words which have a hidden meaning beyond the grasp of the human mind? For who can understand these expressions?» ([3], I, p. 49).

J.-P. Cléro, *Reflections on Medical Ethics*, Philosophy and Medicine 138,
https://doi.org/10.1007/978-3-030-65233-3_3

use of it unavoidable is the issue of abortion: is this not the killing of an embryo or foetus which *could have* become a human being if it had been left to develop without any artificial interruption? There is another case which also comes to mind almost immediately: can a doctor only invoke as a defence that he could not know the consequences of his act when he gave the patient the treatment that killed him – and have agreed therefore to be *held liable* for his acts – but not be found *guilty* of them, according to the famous phrase of a minister of health?[2] What is known at a given moment in the particular case of a patient who is provided with care is not what will be known later. Consequently, in the act of tending to a patient it is understood that what is done to the patient is not always what should have been done, or what would have been done if this had been known. An act could be performed which was revealed to be too timorous, or on the contrary, dangerous or imprudent.

Temporality, intertwined with probability and hope, is essential to the medical act, which is always risky in some degree, however small the risk might be. Taking a risk is knowing that an act which has been committed with the best intention in the world and with all the precautions one was aware of, may end up badly and produce a yatropathy which the patient could have avoided without that blunder, or may even cause his death. Treatments can often be compared to scales in which what one does, has done or will do can reveal itself to be more or less, in the long term, fortunate or unfortunate, without the carer being able to escape the imperative of treating, so that he is as accountable for not intervening as for intervening, with or without the consent of his patient, which makes his responsibility more complicated, without taking it away.

Thus the two previous examples, which are paradigmatic, as it were, show two ways of considering potentiality. On one view, the events we witness and in which we want to intervene are the clues to a substance that gives itself to us only partially, and of which the totality is only latent.[3] This is Descartes' idea, as can be seen in his answer to the second objection of Hobbes against his *Meditations*: 'Il est certain [qu'] aucun accident ou acte ne peut être sans une substance de laquelle il soit l'acte. Mais, d'autant que nous ne connaissons pas la substance par elle-même, mais seulement parce qu'elle est le sujet de quelques actes, il est fort convenable à la raison, et l'usage même le requiert, que nous appelions de divers noms ces substances que nous connaissons être les sujets de plusieurs actes ou accidents entièrement différents, et qu'après cela nous examinions si ces divers noms signifient des choses différentes, ou une seule et même chose'[4] ([3], p. 136–7). This

[2]Mocked by some journalists when it was pronounced, that is, several years after the affair of 'contaminated blood' burst, that sentence of Georgina Dufoix, who had been a minister of social affairs and national solidarity at the peak of the crisis of 'contaminated blood', is far from inadequate. The Court of Justice of the Republic found the minister innocent of the charge of involuntary manslaughter.

[3]That is the terrible avowal of the unknowable nature of what we nonetheless posit at the beginning of what we try to know.

[4]« It is certain that [. . .] no act or accident can exist without a substance for it to belong to. But we do not come to know a substance immediately, through being aware of the substance itself; we come to

conception can be applied to the foetus or to the embryo, which may refer to a humanity which is admittedly not entirely sensitive and present in the phenomenon but which already exists as if veiled, even though the empirical evidence of it is still missing. On the other view, potentiality lies in events likely to announce other, past or future, events, with some degree of probability of the existence of what is announced. There is no substance in the second case, no hidden strength or dynamism, but the simple knowledge – when it exists and can be used – of past experience and experiments.

In a rigorous and maybe somewhat forgotten text entitled *Nécessité ou Contingence* < *Necessity and Contingency*>, and subtitled *L'Aporie de Diodore et les Systèmes Philosophiques* < *The Aporia of Diodorus and the Philosophical Systems*>, J. Vuillemin has shown that since Greek Antiquity, our culture has conceived of potentiality in terms of an antinomy which is somehow related to a previous opposition which seems difficult to solve, though it is not insoluble. The interesting aspect of Vuillemin's work is that he thinks of potentiality as a problem of logic, admittedly as a problem of existential logic, but unconnected with the problem of passionate and emotive aspects which did not wait for 1984 – the date when his book was published – to invade ethical questions and thus prevent their conceptualization. The issue of abortion, which was decriminalized by the Veil Law (17 January 1975) in France, gave rise to passionate feelings in the 80s. The affair of the so-called 'contaminated blood' also sent a wave of indignation and terror through haemophiliac patients, who discovered that instead of being benefitted by the blood they trustingly received, they had contracted from it a fatal illness, and also through the onlookers who, being powerless, and potentially future patients, wondered how could they trust doctors.

While admitting, like Descartes, that passions and emotions give the flavour to life itself, and that, in a way, all of them are good,[5] one should not let them take the place of reasoning, even in ethics, where some constantly remind us that what is at stake is the human being, as if, by trying to reason, to calculate and not to let oneself be overcome by emotion, one appeared as less human. Vuillemin's book is a lesson in humanity and existence, which does not explicitly mention medicine, except for a passing mention of Ammonius,[15] whose name most of us have forgotten, if we ever knew it, but which also makes us suspect, by choosing that allusion, that, despite his silence on the issue, the author is actually constantly thinking about medicine. We are going to wonder if his thesis on potentiality or power which allows him to extricate himself from the antinomy, which we have noticed from the beginning, may have an ethical meaning.

---

know it only through its being the subject of certain acts. Hence it is perfectly reasonable, and indeed sanctioned by usage, for us to use different names for substances which we recognize as being the subjects of quite different acts or accidents. And it is reasonable for us to leave until later the examination of whether these different names signify different things or one and the same thing» ([3], 1985, II, p. 124).

[5]Descartes does not back before hatred and does not say that hatred is never good, as Spinoza, his great dissident reader, said.

3.1 The Dialectics of *Nécessité ou Contingence* (1984)

First of all, what is this thesis, or rather, antithesis, that we are going methodically to state – to begin with – without any direct ethical concern? We intend to measure, in a second part, the ethical scope of the solution that we will have sketched.

There are two possible and contradictory attitudes to a possible future event which may happen, and two ways of considering its probability. The first, which we will call 'realistic' or 'dogmatic', consists in claiming that probability has meaning only for the human mind which considers the event without knowing what it will be, while believing that, even though it is has not yet manifested and thus remains unknown, in itself the event is determined. It claims that, even though it may be impossible to give a phenomenon or an event a degree of probability, it is no less true that it will happen tomorrow – or that it will not happen tomorrow – without any intermediary position. In other words, if we do not know that an event will occur, we can nonetheless produce the hypothesis or the fiction of a sovereign intelligence which would know all that has to be known in a situation and could say whether an event is going to happen, or has happened, before it has happened, or after it has happened, with the same certainty it has in relation to all present events, but without us knowing or having known anything about it, because of our limited intelligence. Probability is not of this world, it does not affect the events or things of this world. It has meaning only in relation to the finitude, failure and puniness of our minds. With apparent modesty, the thesis thus claims that things are what they are, that their probability is due to the fact that we do not know them. The event is or is not: it will be or it will not. It is true that it is or is not, true that it has been or has not been, true that it will be or will not be. But it is not true that it can be both at the same time, as the assignment of a degree of probability seems to imply. Probability does not affect the event. It is not a mode of its being, but a mode of our way of relating to it. What seems to us to be a probability is in reality a power in the thing, one of its properties being to produce the event, or to let itself be moved so that another thing will produce it by its intermediary instrumentality.

To this first thesis is opposed an antithesis which refuses to intelligence the right to project itself onto a point of time of which it would claim that it knows what it does not now know, which therefore refuses it the legitimacy of believing that the event it will know later is already here, in itself, even though it does not know it. From that moment, the situation, as we understand it to be developing, must be taken into account in the entirety of what is given to us – which does not exclude our having to look for what is given to us – and in such a way that its becoming cannot but be calculated. We will only consider its becoming – its chance of occurring in the future or its chance of having occurred in the past – in a probabilistic mode which is the very being of that situation. In other words, the mind does not unload itself on a being that it imagines knowing in its place – through some imaginary delegation – that which it does not know. It does not suppose an underside of the cards, hidden by some veil in some substance about which it does not know anything. There is no veil. The situation presents itself as being what it is. The actor who seizes the situation produces his information, projects acts from it, defines strategies, starts acting to change it, and presents himself as a partner of the circumstances he confronts. This is far from the first thesis which views him as guessing, revealing, discovering or interpreting a situation which would have an underside, which should be imagined as escaping him, which could be known in itself, but not by him. According to this antithesis by contrast, probability is therefore a sort of specific mode of being which it is impossible to overcome unless by fantasy.[6]

The pride of those who support the thesis, who have put on a feigned modesty, is that they think they can imagine a transcendent being which, under the form of God or the Laplacian intelligence, knows in their stead, without them having or being able to have the slightest idea of who that transcendence is or of what it knows. It is but too easy to argue against them that one should be wise and keep to what one can know – without faking a scenario of which one has no representation – by behaving as if one knew since somebody – of whom one neither knows or wishes to know anything – is supposed to have perfect knowledge. Conversely, the supporters of the thesis gleefully accuse the supporters of the antithesis of making man the measure of everything, of acting as if man had no 'other', and as if everything that constituted situations was only what appeared or was discoverable by his intelligence. The supporters of the antithesis are thus held guilty of an even worse pride than that of which the proponents of the thesis are accused: while the latter may be accused of imagining a place for the transcendent, the former can be accused of usurping it.

In this reciprocal accusation of the two sides of the antinomy, which should be considered to be right? Should we be on the side of those who think that there is an underside to what we know of situations, that it should be calculated, guessed, interpreted, submitted to a sort of ἀλήθεια? Or, should man, in a Promethean conception of himself and of things, set himself as a key actor in all situations, not in the subordinate role of interpreting them, but in that of assuming them without knowing, it is true, if what he does is the best thing to do, or if other actions could have been possible with superior effects bringing more pleasure, more happiness to actors? *In the first case*, the event is understood as a clue or indicator of a substance or strength of which one allows oneself to suppose that it mostly escapes one and already contains all the future events; *in the second case*, the event does not reveal any hidden substance or force, but only indicates other events that one anticipates or considers as having already taken place. In the first case, we could talk of

[6]Vuillemin, who uses history, especially ancient history, and gives extraordinary depth to the theatre of the reflections to which he leads us, finds the origin of that attitude in Aristotle himself who, though he did not calculate probabilities, nor even thought of doing so, thought that, in an alternative of statements about the future, what is true and necessary is not one of the other components of the alternative but the latter itself, which is really constitutive of a being which is neither that of an event nor of its absence. That particularly emphasizes the idea of power and radically thwarts the necessitarianist illusion of the Megarics.

determinism; in the second, of *subjective probability* or of *Bayesianism*,[7] which results in the fact that in any situation I calculate acts to be performed, that is, the production of other events based on the events I have at my disposal. Thus, I can, in any given situation, assess the probability of its evolution, or assess the probability of its origin, with chances of being right, or wrong, which the calculation determines. I project my activity onto a situation and measure its chances of succeeding.

It is clear that, in this antinomy, the first thesis is burdened with a supposed knowledge that cannot be guaranteed at all, for there is no possibility of placing ourselves outside events and phenomena in order to know how things are 'in themselves'. At the moment when I do not know something and I cannot know it, though I have tried to know it with all the means at my disposal, it is useless to suppose that someone knows it or can know it, since I only desperately pretend to fill up a place that remains empty. On the contrary, I am not feigning any knowledge when I keep to the phenomena and events I have at my disposal, to those I – or others – have already experienced,[8] and to the assumptions that I take for nothing else than hypotheses.[9] In *the first case*, I call *necessity* a process supposedly internal to things themselves; in *the other*, necessity is the very high probability that an event or a phenomenon will occur or has occurred. It is quite obvious that the antithesis is more 'activist' than the thesis. The latter subordinates action to a supposed knowledge which only exists in fantasy. The antithesis, on the contrary, puts available knowledge to service in satisfying the demands of action, which is given supremacy.

We have reached the threshold of our difficulties, since the problem which Vuillemin leaves us facing, without confronting it himself, which is that of *potentiality in ethics*, has two main aspects. The *first* is to know whether it is possible to find a carer guilty of an action when he took into account all that he could of the given circumstances, the development of which has shown that he did not take the right decision for the patient, that he was thus wrong when he took it. The *second* is to know if a being which is at the beginning of its development, and which has very high chances of reaching its full term, should already be considered now – even though its development is not finished – *as if* it contained its completion, and whether, consequently, it must be accorded the same rights as the finished being it

[7]From Thomas Bayes, the eighteenth century theologian-mathematician, who invented them in a small essay published in London in 1763 thanks to R. Price and entitled *An Essay towards solving a Problem in the Doctrine of Chances* [1].

[8]And of which I – or others – have already experienced the developing effects in the shape of other phenomena or events.

[9]I wish to underline here that, contrary to what might be imagined, this point does not have much in common with religious belief. What are called 'subjective probabilities' or Bayesian probabilities were invented by Thomas Bayes, who was not only an eminent mathematician who elaborated the Rule that bears his name, but who was also a theologian of the Presbyterian Church, a dissident Church of Anglicanism. Christianity does not at all imply support for what I have called dogmatism. Conversely, what I have called determinism is not only the work of theists. Atheists could also suppose some necessity in the processes of matter which would not be due to some superior intelligence organizing them.

is not, but which one presumes it will become, or similar rights which hold that completion to be established.

3.2 Points on Which the Bayesian Theory of Probabilities Triumphs (Supported by the Antithesis)

What we have just called the first aspect of potentiality does not seem very difficult to solve, at least in principle. One could not reproach someone with being wrong if, at the moment when he calculated what could happen in given circumstances in consequence of his acting as he did, he took into account *all* the aspects he should have taken into account. He can be considered to have made a mistake in the light of additional information related to the later development of the situation, but mistake and truth have no meaning except in relation to given circumstances. A doctor who can justify his way of considering a situation, and who took a decision which later was revealed as ineffective or even disastrous for the health or life of a patient, has not been wrong, properly speaking, and has not necessarily chosen a bad solution, even though it was eventually disastrous. When, under the pressure of the pharmacist Homais, health officer Charles Bovary operates on a club foot and fails in his operation, he is to blame, because he has not measured his ability correctly and has set about doing something which had every chance of going wrong, but when one does not know the dangers of HIV, does not suspect that it is present in the blood with which one is inoculating people who need it, there is no mistake on the part of the doctor, even if he can be held liable. New data may emerge in the light of which his decision will turn out to have been wrong, but he was not in a situation of being able to know them at the moment when he intervened in the given circumstances, nor even of calculating them as they appeared.

In so far as things as they are in themselves escape us, and we can neither read into the future nor into a past which leaves few traces, one should talk of truth and mistake only in relation to circumstances. One can, by considering the same events, say that a carer has not taken a wrong decision, at the moment when he decided upon a treatment, even though, with all sorts of extra information on the same events, one is later in a position to assert that the decision was not a good one and that it could have been better. Difficult though it may be to hear and admit for those who are discomforted by relativism, the true and the false are related to circumstances. What is true or adequate in some circumstances may be wrong or inadequate when those circumstances develop so that we have other information which radically changes them. The information we have is not outside circumstances: it is intrinsically part of them and constitutes them. One could then push the paradox further and say that someone who, with hindsight, and from the point of view that new knowledge of those circumstances now offers, finds themselves to have taken a good decision, may not have taken one at the moment when they took it, and that they were simply lucky that it was the best possible one, though they could not do anything about it. One could say that the generalization of the use of oral contraceptives in the 1960s was a good decision, but, at the same time, that doctors were lucky when they prescribed them, for it was not beyond possibility that they would trigger all sorts of illnesses in the patients who were prescribed them. Seen from 2019, that is, more than a half century later, it is not true that that generalization was carried out then with all the required prudence, even though it was an excellent thing that it was practised thanks to the doctors' taking some distance from the precautionary principle.

Thus the first practical consequence – which is linked to the fact that one cannot say that someone was wrong in given circumstances from the perspective of a truth that emerged later – is prudence in laying charges against carers, since, being obliged to treat, they have committed no negligence either in relation to the knowledge required or in the collection of the information necessary to take a decision. If the doctor is always liable for what he does to his patient, the latter must also be aware that he asks to be treated at the moment of his illness, and that the treatments to which he is submitted for his recovery, or in order for him to get better, though they are plausible at the moment when they are prescribed, may not be thought to have been the best possible later, without his being obliged to incriminate his carer. One virtue of the patient is that of acting in full consciousness of this parameter, and not to overburden those whose duty it was to help – even though he might have to bear the consequences – and who loyally tried to do it. The critical point is obviously the question of knowing whether the information could have been fuller or more relevant at the time, and if the level of competence of whoever was in charge of treatment was high enough. Provided those two points are fulfilled, the respect of Bayes's rule, which gives he who acts the leeway they need, tends to exonerate a practice which was later revealed to be defective.

3.3 Nonetheless Meets an Obstacle

The most difficult case may be the one which we noted in our introduction as the second aspect of potentiality. For it is possible to be satisfied with drawing a distinction between a *de re* potentiality and a *de dicto*[10] one, and to be content with the latter to solve the previous problems, or at least some of them, and certainly those which, up to a certain point, are about only the relation of truth and the statements made in given situations about events that are considered from two (or several) different temporal viewpoints. In the cases we are going now to consider, and of which abortion is the paradigm, it is difficult to get out of trouble thanks to the distinction between *de re* and *de dicto* because the potentiality of the beings we are talking about, when the issue is abortion, for example, is not only one of speech but also a real fact. One can, in other words, encounter the same problem at one remove. Although, in the cases studied in the previous chapter, in particular in psychiatry, it is

[10] *de re*: about things; *de dicto*: about discourses.

possible to be satisfied with considering discourses and the different points of view from which they are conducted, in the cases which we are going to examine now it seems no longer possible to do otherwise than consider the beings themselves, the demise of which is allowed or the killing of which is forbidden. In still other words, the mode of being which, from Aristotle to Savage, by way of Pascal and Bayes, has proven its worth against ontological dogmatism – that mode of being by which the probable gains offered by some sort of autonomy – may well be countered. The thought that being is and not being is not, without any intermediary being granted semi-rights to live and only likely to have a half-death, may impose itself in some cases. Indeed, are there not cases in which practice itself invalidates the probabilism by means of which one has tried to vanquish and has hoped to vanquish once and for all a dogmatism which imagines real beings and real positions behind what appears to us? In the end, should not those *real* beings and *real* positions be considered, despite our *theoretical* reluctance to pose them in themselves? Is the autonomation of the *being for the possible* or of the *being for the probable*, which Vuillemin noticed in Greek Antiquity and which seems so solid against dogmatism, not doing violence to reality, when the issue is one of judging situations which involve the right to abort? Despite its ontological phantasmagoria, would not dogmatism be right after all, at least in the intention it points to, of touching reality beyond the game of representations and the calculations they produce? One understands that this issue should be looked at more closely, and, from that point of view, that abortion, which is undoubtedly the most crucial and symptomatic instance of it, requires further elaboration.

Here our consideration must take into account Peter Singer's point of view for, not without a degree of imprudence, he has declared that he could put an end to the controversy by reducing dogmatism to silence: 'In contrast to the common opinion that the moral question about abortion is a dilemma with no solution, I shall show that, at least within the bounds of non-religious ethics, there is a clear-cut answer and those who take a different view are simply mistaken.' [11] Singer reduces to a syllogism the argument at the end of which abortion indisputably appears to be a murder. Supposing one gives some importance to *Thou shalt not kill*, an order which leaves its object completely without any determination – what or whom do I have any right to kill? – one should acknowledge defeat by the following reasoning:

1. One must not kill an innocent human being. (First premise)
2. An embryo or a foetus is an innocent human being. (Second premise)
3. So, one must not kill an embryo or a foetus. (Conclusion)

Everybody will understand that the weak point of the argument lies in the second proposition, which, beyond *Thou shalt not kill* – the imperative of which, abruptly stated, seems to set us on an equal footing with things themselves – makes us return once more to the previous antinomy, which, under the form of a simple assertion, opposes an ontological dogmatism to a critical cognitive point of view which we have deemed more acceptable. Ontological dogmatism easily takes the form of an assertion of the continuity of existence from the undetermined state of a few cells – or even the fusion of two gametes – up to the being the human attributes of which are quite indisputable (human form, presence of pleasure and pain, self-consciousness, for example), as if, given a single impetus, one went from one stage to the other – even though that impetus lasted several months or years. The critical and cognitive point of view insists, on the contrary, not only on the difficulty of assigning human characteristics that might be decisive, but also on the lack of certainty one can, and must, grant to the reality of continuity, which is at the heart of the thesis. The thesis, which asserts the continuity between the foetus, and even from the embryo, and the complete human being, as if it were one being, is but an ideal and fictitious reconstruction that one produces by looking back from the (estimated or projected) state of completion and going back to the origins. After that, once those origins have supposedly been reached by thought, one pretends one believes that existence has undertaken the reverse course which one thinks of as real, and not only as a simple hypothesis. Of course, the ethical transcription of that metaphysical idea consists in saying that, since one does not grant oneself the right to kill a child – which is the most common ethical option – one should not, because of the principle of continuity, kill a foetus, or even an embryo. It is easy to denounce in the antithesis the false value of a reality that is given in the thesis by using – implicitly or explicitly – the idea of a *potential humanity* which would permit us to treat as the same being the being that will be and the being that is already now – as if the first was contained in the second, as a property of it. That idea of *potentiality* is a fictitious entity meant to hide the reversal we have just mentioned, to treat as identical to a foetus a being which seems scattered by its becoming into a myriad of fragments and events, and to give oneself the famous ontological minor premise, that *an embryo or a foetus is an innocent human being*. The idea of *continuity* used as a principle[11] and in two directions – back to the origins then forward from them – is indeed much less clear than it seems. The two movements it claims it is identifying are not of the same essence, and the one does not make it possible to establish that the other is well founded. Thus, as Kierkegaard noted, 'the past is not necessary, in as much as it came into existence; it did not become necessary by coming into existence (a contradiction), and it becomes even less necessary through any [ontological] apprehension of it.' [7] And the set of 'possibility(ies) from which emerged the possible that becomes the actual always accompanies that which came into existence and remains with the past, even though centuries lie between.' [7] A contingent event does not become necessary because it has moved into the past, even though it is no longer possible to change it. Unchangeability does not prove necessity.

Similarly, from which point of view could continuity be asserted to guarantee that a being belongs to mankind – whatever the level of its development – apart from a fantasized point of view we will never have? Even though the awareness we have of ourselves does not guarantee its value, we will not escape the projection that we are the same, towards either the past or the future – the continuity of that identity in its becoming being the very definition of potentiality. What we remember as constituting ourselves is certainly not ourselves, nor is what we consider as being ourselves.

[11] We will have the opportunity of criticizing it again, but from another, ethical, angle in Chapter V.

There again arises the opposition between, on the one hand, the claim of saying things as they are, even though we only fantasize them, and, on the other hand, the methodical point of view which makes sure it does not leap from the conditions of knowledge to the conditions of existence, as if they were the same thing. They are so little the same thing that, if one refuses to grant that the identification be realized by an unacceptable coup de force, it must be calculated by granting it only degrees of probability. Have we won the game? Not yet.

Even though, following the wise advice of Nietzsche, we were reluctant to oppose theory and practice [9], we would nonetheless be compelled to admit that practice sets the issue in a much more acute way than theory. Supposing that I am a doctor, quite convinced by the reasons of the antithesis, and a supporter of the critical viewpoint of the method, if one of my patients asks for an abortion, am I going to perform it as if the being that is to be killed had not much more importance than an appendix, or a subordinate and defective organ of which the patient should be relieved? This is, as we know, one of the pieces of the analogical argument used by Judith Jarvis Thomson in favour of abortion [10, 13]. Does not believing that one is 'already' dealing with a human being bear, at the moment when the carer is asked such a thing, a good part of the value that the critical argument disputes, when it tries to create a specific category for the beings which are only potentially human, understanding that they may only probably ever be so? In other words, does not action, in its Manichean nature – either you do it or you do not, which excludes any third position – give an advantage to the thesis that, for all its defects, we have called *dogmatic*? Is not the antithesis – despite all its theoretical virtues – trapped in its own subtlety, and rendered quite incapable of giving a practical interpretation of its willingness to create a sort of autonomy of status for the probable situation, for *either* the foetus is left to live *or* it is not, without any in-between being possible? We admit the incredibly paradoxical character of that philosophical situation, since the thesis, which is more contemplative than active, is nonetheless granted the ultimate point of reference for practice because it does not deny that there is a reality in itself, whereas the antithesis, which is from the beginning more practical than theoretical, is being denied – at the moment it was about to triumph on a razor's edge – the only confirmation that it lacked, that of a contact – even a fugitive one – with reality. In some situations, therefore, in particular in those in which life is at stake, it seems unavoidable that practical supremacy be granted to 'dogmatic' positions, precisely because, in them, the necessity of choosing between one option and another disrupts or disqualifies any probability. As soon as a situation does not leave any room between an option and the opposite one, it seems that Bayesian probabilities lose their credit and leave the theoretical attitude informed by them in an untenable position, a position too fragile, in any case, for a practice which does not support them because they do not enlighten it.

Admittedly, Singer claims he is able to do without probabilities, and thinks he has found the solution in favour of abortion when he says that eliminating a foetus – and even more so an embryo – is not more nor less important than eliminating an animal that did not have any consciousness of itself. He is wrong, however, because when he says that, he shows as little knowledge of probabilities as the opposite point of view, which despises that knowledge, though it does not sever the situation or the being that is in question from its becoming. Singer, for his part, separates the situation from its becoming and arrests it at the stage of what it is in the present, in a way exactly parallel to that in which his adversaries do not take any account of the developments of a situation the dimensions of which are nonetheless obviously probable. He does not take into account the concept of *potentiality* as probability any more than they do, even though he recommends acting in the opposite direction to that recommended by the dogmatist, who makes no distinction between an abortion and a murder, especially an infanticide. When Singer writes, 'My suggestion is that we accord a life of a fetus no greater value than the life of a non-human animal at a similar level of rationality, self-consciousness, awareness, capacity to feel, etc. Since no fetus is a person, no fetus has the same claim to life as a person,' [11] his suggestion overlooks important features and should be rejected. In addition to the argument being seriously flawed,[12] Singer makes a substantive choice, contrary to his promise of reducing his adversaries to positions that are clearly mistaken. When he draws the consequence that it is no more no less forbidden to put an end to a 'human' (humanoid?) existence by abortion than to kill an animal, since that existence does not have more inherent value, he conceals a parameter that is too important to be granted without examination at the most provocative part of his argument. It is not true that an abortion should be compared to the sacrifice of an animal.

When one wants to operate the antithesis like Singer, positively rather than apagogically, it sounds out of tune and only provokes cynicism. That a sort of Hegelianism is used to classify animals depending on the self-consciousness they exhibit, and that the strength of their right not to be eaten is inversely proportional to their proximity to the fully self-conscious end of a continuum running from unconscious to self-conscious is one thing,[13] but that the stages of the development of man are inserted in the interstices left by the spacing out of animals according to the level of self-consciousness they have [11] is violent and unacceptable, and only reflects the positions that we have called *dogmatic*, and the use of transcendence and the unavoidable speciesism which we, along with Singer himself, contest. Indeed, Singer's philosophy radically avoids speciesism, but its rationality becomes dubious insofar as it is but too visible that, in one of his arguments, it originates in an arbitrary separation or severing which claims to arrest the living humanoid at the stage where it is, and deliberately denies the development that, thanks to the idea of probability, it must nonetheless be granted. What Singer understands as the 'inherent value' of a

[12]Though the foetus is not a person, that does not mean that it does not have any right to life. It may not have one as a person, but that does not preclude other rights to life, unless only persons have a right to life (which is absurd), or there is no other right to life than those which result from being a person.

[13]Hegel used spacing out according to self-consciousness to classify peoples. In relation to the diversity of species, that criterion becomes quite arbitrary. Each species, if that unit is accepted, presents a temporarily balanced set of features which allows it to live. Isolating one, to the detriment of the others, is something that is perfectly arbitrary.

life is that life is frozen at what it is in an ordinary present – without any taking into account of the probability that it could have become a very superior life under the aspects, which he himself lists, of self-consciousness, rationality, capacity to feel, etc. – which makes his position weak at the very moment when it is expected to be at its strongest, since it also appears to be at its most scandalous and provocative.[14]

This is all the more the case since R.M. Hare, his master at Oxford, had shown, in an important chapter, [4] that the utilitarian calculus is not necessarily in favour of abortion in all possible cases. In other words, there can equally be bad reasons to abort. His arguments are not always very convincing, and he seems to mix acceptable points of view with others which are quite unacceptable, but he nonetheless highlights that if the happiness of the mother can legitimate abortion, it is not contrary to a calculating reason to say that, by killing an embryo or a foetus, one kills a being which could, as fully as myself, have enjoyed life and the happiness of living, without that being having the means to reject the act that kills it, and which, upon reaching self-consciousness and enjoying a happy life, and looking back later upon this act would probably have rejected. Hare admittedly leads us through very high degree fictions. That is probably what earned him the attack, which was only implicit, as Singer did not directly target his former master in *Practical Ethics*, where he seems to proceed in a more hidden manner, letting Tooley, whose famous article he edited in *Applied Ethics*, say, 'The retrospective attribution of an interest in living to the infant is a mistake [10]. I am not the infant from whom I developed. The infant could not look forward to developing into the kind of being I am, or even into any intermediate being, between the being I am and the infant. I cannot even recall being the infant; there are no mental links between us.' [11, 14]. Hare makes a few swerves in the pseudo-logic of things in themselves.[15] Hare's argument, which extends in argumentative time an episode that lasts only for a moment and has meaning only during that moment, is thereby quite weighed down. He nonetheless indisputably puts a weight in the scales where fictions are to be weighed which is far from being insignificant.

In *Essays on Bioethics*, Hare sets the issue of abortion in the following terms: 'En considérant la mise au monde d'une nouvelle personne, je dois poser la question comme si je devais être cette personne ou comme si cette personne devait être

[14]To be quite fair, one must add that just after the developments I have just mentioned and which he complacently spreads out, he analyses a syllogism that holds that the minor premise (the second proposition) is stronger than that of the previous syllogism. Here is the syllogism in its entirety:

First premise: It is not right to kill a potential human being.
Second premise: A human foetus is a potential human being.
Conclusion: Therefore it is not right to kill a human foetus.

The weakness of the syllogism this time, unlike the one we have previously presented, is due to the poverty of the first premise: it is not enough to be potentially a being to have all the rights of that being. Does Prince Charles have the rights of a king of England, because he has had the title potentially for a long time?

[15]This tends to confirm what we have just established: to judge abortion, it is difficult to make do completely without dogmatism.

moi-même. Ainsi deviendrait pertinent ce à quoi, pour cette personne, cela ressemble d'exister; et, évidemment, ce que cela lui ferait de ne pas exister.' [4] The idea of the *happiness of being born* is mentioned later [4]. As to the detail of the argumentation, it becomes apparent when he supposes that 'it was my [his] own mother deciding whether or not to terminate the pregnancy which actually resulted in *me* [him]'. [4] He then tries to weigh, from the point of view of what exists, the deprivation of existence that an abortion would have constituted for the foetus at the stage at which it was, and to weigh the case for an abortion carried out on the foetus which has, in the event, developed into the being he now is: 'The question is, "What *do* I (a presently existing person) now say about this past situation?". I will draw attention to an obvious reason why I might not like to say that it was all right for her to have an abortion. The reason is that if she had had an abortion, I would not now have existed. Let us suppose that I am able to reach back in time and give instructions to my mother as to what she should do. Suppose, even, that she is able to ask me questions about what she ought to do. In order to get into a position in which I can communicate with her at that time, I shall have to penetrate some noumenal world outside time (this is really getting very Kantian) and have access to her in that past time. This of course raises deep philosophical problems, into which I am not going to go. But just I suppose I can do it. What shall I say to her? I am sure I shall not say, "Carry on, have the abortion; it's all the same to me." Because my existence now is valuable to me. I shall not, other things being equal, will (. . .) that she should have the abortion, thereby depriving me of the possibility of existence.' [4].

Despite what Singer says, when the act of killing is at stake, it upsets all the scales upon which one would like to measure it. Though the mode of argumentation of the dogmatic is detestable and pushes a rational mind towards the critical side, one must admit that the latter becomes incredibly fragile at the moment when one must decide about a life, be it called *human*, or *quasi-human*, *already human* or *not yet human*. Admittedly, the legal argument that we have used in the case of an unequal relation to the true and the good, that time allows – depending on whether we are situated before the event or after it – is still true here, since jurists wonder about the status that could be granted to a foetus which dies before it is able to become a new-born.[16] [12] When the question is about taking a life in conditions in which it could have been saved without inflicting any damage on other lives, it seems that the moment one decides to kill does has different properties from the moments of deliberation before the event or after it. The moment of the decision is not related in the same way to the calculation or assessment of the degrees of probabilities as the moments that supposedly prepare the deliberation or draw the conclusions of what has been achieved during a discussion. The case of an execution is of course crucial. Singer has seen this well enough, as he has written many pages on precisely what he thinks is the overvaluation of human life, or in his words <'the widespread acceptance of

[16] Claude Sureau's article is available freely on the Internet. Without recommending that the embryo and the foetus be treated as persons, it nonetheless criticizes the lack of status attributed accorded to the unborn child.

the doctrine of the sanctity of human life'>, which is the very nub of the issue. However he mentions too quickly the taboo in regard to killing, or the sacred and irrational character which surrounds the act of killing, as if it were a simply a hindrance to reason, which ought not to be obliged to endure such hindrance. But why should there not be a rationality which, while upsetting the Bayesian scales we know, would have its own laws, which we would have to build? And why would those laws prescribe in exactly the same way close to an event as very far away from it? Could not a defensible rationality, rather than dismissing all those differences of assessment which depend on the supposed closeness or remoteness of an event, rather consist in studying their mode of operation?

One might incidentally wonder whether, overcoming the 'tragic' character of the execution, the same does not apply to all the decisions which, as soon as they are important, seem to smash to pieces the logic of deliberations.[17] This does not mean that it is not important to deliberate, but it does mean that that the decision cannot be deduced from the deliberation and is not one of its outcomes which is of the same nature as the rest. There is, in the decision, a moment of reality which defies any rationality of calculus.[18] That rationality may question the moment of reality and find it illusory. It may be illusory, or it may not be. As we cannot know, we cannot but concede an agreement with reality that deliberation never reaches, because it remains a discourse, removed to some distance, whatever it may be, from reality. It may be that, in the short moment when the decision is taken, in which it comes into contact with reality and starts a discourse with it, the decision does not owe much to the deliberation and to the game of representations with which we know how to deal. Ethics has meaning only in those moments of decision which set it on an equal footing with reality, even if only thanks to an illusion. We do not mean that criticism must settle on the ethical field in the most demanding way, but reason can only get a grip from that angle. Criticism must acknowledge that it is incapable of reaching the moment of existence itself, the moment of the act which is between that in which it still remains to be performed and that in which it is not to be performed anymore because it has already been performed. This moment of upsurge is a Zenonian one for it: either it is past or it is still to come. The actual passing escapes it, and when one tries to give it a rational form, it does not have to hand much else than dogmatism, which is so easily refutable in all the conditions of deliberation. A philosophy may be bad in deliberation and true only for one moment – that of the discourse where the symbolic seems to meet the 'other' – as another philosophy can provide an excellent instrument for deliberation by letting the decision, which is almost the only point of action which counts, completely escape.

[17]The archetype of that collusion of the instant of death and the instant of decision is, obviously for Kierkegaard, what Christianity has called 'Abraham's sacrifice'.

[18]Is it by chance that in *Philosophical Fragments*, the noun instant is almost constantly associated with the noun of decision or with the adjective decisive?

3.4 Conclusion

In so far as no truth exists without being temporal, a *chronicle* or a *chronology* of practical discourses – since they are not all right at the same time – should be written. The issue is not to disqualify them from the point of view of reason, but to seize the moment when they are true. The degree of proximity of reality – even for a problematic case – implies extremely different assessments depending on the time when they are made. The strength of Bayesianism is to limit and surround the moment of the decision, to give it a place and a horizon. Its weakness is to leave a gaping hole at the moment of the decision, while denying that it is uncontrollable. Dogmatism, which is almost always false or illusory, can only inform philosophy for a very brief moment of practice, and can be right only during that very short moment, leaving the field clear for criticism the rest of the time. In an article of Humean inspiration,[19] [2] we suggested the idea of the truth of the philosophical systems as depending on the time distance supposed by the actor at the time of his death. While keeping the 'chronological' idea, this formula should be amended, for the mortal event is not only that of the thinker who takes on responsibility for a philosophy. It is also, in some situations, that of a being who is interested in the issue of his life or death, which diversifies and refines the scope of the question very much. We know now, however, that it must be widened to include the act of deciding. That truth is temporal is certainly taken into account by Bayesianism, which we have fully considered in our antitheses, but we have had to deprive that Bayesianism of its last encounter with the moment when the decision meets reality, ceasing for that short moment to be symbolic and finding itself symbolic once again just afterwards, even though under different conditions than before, and again subjected to the Bayesian rule, since for the moment there is no other ethical rationality of medical practice which grants temporality so much attention. The antinomy has, therefore, opened the door to a task that would be that of a *chronicle* in action-philosophy or ethics. If one wanted to see some kinship with the contemporary theory of stories or story-telling, we would have no objection, provided that the holes in that story or tale are as important, and even more important, than the fullness of events which are told in the representative mode.

The philosophy which would accept that the thesis could be philosophically valid only for a brief time, and that the antithesis would be valid the rest of the time, has not been written yet, except perhaps by Kierkegaard, who has thought of the *instant* by indicating its place beyond the reach of Greek philosophy, which excluded it from the category of the thinkable: 'If, then, *the moment* is to have decisive significance – and if not, we speak only Socratically, no matter what we say, even though we use many and strange words, even though in our failure to understand

[19]In addition to discourses on the temporality of our passions and on the relative liveliness of our impressions and ideas, Hume is the author of four discourses which present four philosophers – the Platonist, the Stoic, the Epicurean and the Sceptic – who each have their 'moment' of reason [6]. This has sometimes been called the picture of the four discourses, but we prefer calling it a quadrant.

ourselves we suppose we have gone beyond that simple wise man who uncompromisingly distinguished between the god, man and himself, [...] then the break has occurred, and the person can no longer come back and will find no pleasure in recollecting what remembrance wants to bring him in recollection, and even less will he by his own power be capable of drawing the god over to his side again.' [7] And, later on, 'Whereas the Greek pathos focuses on recollection,[20] the pathos of our project focuses on the moment, and no wonder, for is it not an exceedingly pathos-filled matter to come into existence from the state of «not to be».' [7] It was also Kierkegaard who thought of establishing the instant as a requirement for thinking about ethics: 'If the situation is to be different [i.e. if the temporal point must be something other than a nothing], then the moment in time must have such decisive significance that for no moment will I be able to forget it, neither in time nor in eternity [...].' [8] He tried to meet this requirement, especially in *Philosophical Fragments*.[21] The rational possibility of the thesis could then, paradoxically enough, be its best rationale for describing what happens for a brief period of time. The 'completely different' that reality is, when one touches it, could, for that very short time, have its own philosophy of a sort like that of the thesis, which is quite incapable of applying beyond that instant when it claims it can go further than any representation. The difficulty of the antithesis, even though it is cleverer than the thesis, is that in the representations it deploys to surround and embed the moment, it lets it slip, as if through a hole. This is the reversal of the point made by Kierkegaard: 'Pour qu'un point de départ temporel ne soit pas du néant, il faut que l'instant dans le temps ait une importance décisive.'[22] [8].

Thus, the antinomy which was our starting point and which we borrowed from J. Vuillemin is not defeated, but it has transformed itself and has become the following: the dogmatism of the thesis does not describe the requirement of reality well, and the antithesis should ordinarily be preferred for its methodical superiority, but the antithesis lets the thing that it itself knows is the most important thing slip away, and it even seems to have been built to let it slip away. The situation therefore is, quite paradoxically – insofar as the philosophy of the moment has not been written yet and we do not even know whether it is possible – to take as a provisional morality[23] that of the calculations that we have called Bayesian, and to find the

[20]He could have added: 'And that modern pathos concentrates on the future.' We are also thinking of the philosophies of Fichte, Hegel and perhaps, before them, Kant, whose practical philosophy may be read as a promotion of the future.

[21]It is, from this point of view, quite impressive to compare the way Hare thinks about the set of representations surrounding the issue of birth and abortion, and that in which Kierkegaard tried to think in *Philosophical Fragments*, and in a context which is not that of the thinking about abortion, nor even about birth from the instant. [7]

[22]To put it clearly: has the importance that a decision can have. *Riens philosophiques* [8].

[23]We borrow the expression *provisional morality* from Descartes, but everybody will have understood that with these words we are absolutely not defending the content that he gives to his provisional morals, the principles of which turn their back on the probabilism we have defended, and which is closer to game theory.

symbols and acts which suit a philosophy of the moment – which we know remain to be invented, but of which for the moment we have only intuitions – and the symbolism of religious discourse.

Lastly, the opposition that we find in some cases between what is decided at a real moment and what is calculated by playing on the fantasies of criticism may be only a category of what R.M. Hare has thematized in *Moral Thinking* by distinguishing what he calls *intuition*, which works *prima facie*, and critical thinking. Hare does not systematically agree with critical thinking, which is more playful than the intuitive thought which guarantees the seriousness of existence, without being capable of giving a valid reason for it.[24] Hare's philosophy, which we examined at length in the first volume, may very well have to be deeply thought anew according to the terms of a philosophy of fictions, which puts the attachment to reality into perspective, multiplies the critical levels, and above all does not leave the intuitive level as the only one able to claim contact with reality. The intuitive level is not the ultimate basis on which the symbolical or critical levels are supposedly built. It is never more than an initial symbolic level which is liable to be overcome, even as it itself has overcome other levels which seemed intuitive to it, and which have been forgotten as having been purely symbolic. It simply gives the impression of being more real and less symbolic than the other levels because we have become used to it, and because it seems less difficult to seize it than other symbolic levels which require more effort.

We are going to check that operation by investigating the case of blood transfusions which gives rise to strange ethical assessments – and certainly to the paralogisms we are going to examine – but also to an attitude that seems characteristic of ethics: namely that what was acquired by criticism cannot necessarily be immediately monetized into acts. The temporality and the logic of reality are not those of its symbolic systems.

Bibliography

1. Bayes, Th. (2017). Essai en vue de résoudre un problème de la doctrine des chances, Hermann, Paris; Bayes Th., *An Essay towards solving a Problem in the Doctrine of Chances*, in *Essai en vue de résoudre un problème de la doctrine des chances*, Cahiers d'Histoire et de Philosophie des Sciences, N° 18, – 1988, Société Française d'Histoire des Sciences.
2. Cléro, J. P. (2016). Memento mori ou: Comment l'éthique, qui est une pensée de la vie, peut-elle inclure l'idée de la mort? *Ethics, Medicine and Public Health, 2*(2), 246–255.
3. Descartes, R. (2010). *Règles pour la direction de l'esprit*, règle XII, in: *Oeuvres philosophiques*, T. I (1618–1637). Paris: Garnier, 2010. Translation in: *The Philosophical Writings of Descartes*, trans. J. Cottingham, R. Stoothoff, D. Murdoch, Cambridge University Press, Cambridge, London, New York, New Rochelle, Melbourne, Sidney, 1985, vol. I, p. 49; Descartes, René, *Oeuvres, Méditations et Principes*, vol. IX, Vrin, Paris, 1996, pp. 136–137. *Objections and Replies*, Third set of objections with replies, in *The Philosophical Writings of Descartes*,

[24]We are especially thinking of what is been established at paragraph 8.3 of Chapter VIII of *Moral Thinking*. [5]

trans. J. Cottingham, R. Stoothoff, D. Murdoch, Cambridge University Press, Cambridge, Cambridge, London, New York, New Rochelle, Melbourne, Sidney, 1985, vol. II, p. 124.
4. Hare, R. M. (1993). *Essays on bioethics* (p. 72, 173, 154). Oxford: Oxford University Press.
5. Hare, R. M. (1992). *Moral thinking, its level, method and point*. Oxford: Clarendon Press.
6. Hume, D. (1742). *Essays, Moral and Political*, vol. 2. Edinburgh: Alexander Kincaid.
7. Kierkegaard, S. (1985). Philosophical fragments. In *Kierkegaard's Writings*, VII (H. V. Hong & E. Hong, Ed., Trans.). (pp. 13, 18, 19, 20, 21, 79, 86). Princeton: Princeton University Press.
8. Kierkegaard, S. (1948). *Riens philosophiques* (trans: Ferlov, K & de Gateau, J.). Paris: Gallimard NRF, p. 62.
9. Nietzsche, F. *La Volonté de Puissance*, T. 2, (trans: Albert, H.). § 256, books.Google.fr/books?id=HOswDAAAQBAJ&pg=PT21332Ipg=PT2133&dq=Nietzsche
10. Singer, P. (Ed.). (1986). *Applied ethics*. Oxford/New York: Oxford University Press.
11. Singer, P. (1993). *Practical ethics* (2nd ed., p. 97, 137, 150, 151). Cambridge/New York/Melbourne: Cambridge University Press.
12. Sureau, C. (2006). L'inconnu dans la maison. In *Cités* (n° 28) (p. 25). Paris: PUF.
13. Jarvis, T. J. (1986). A Defense of abortion. In Singer (Ed.), *Applied ethics* (pp. 37–56). Oxford/New York: Oxford University Press.
14. Tooley, M. (1986). Abortion and infanticide. In Singer (Ed.), *Applied ethics* (pp. 57–85). Oxford/New York: Oxford University Press.
15. Vuillemin, J. (1984). *Nécessité ou contingence* (p. 158). Paris: Les Éditions de Minuit.

Chapter 4
Blood – One of the Most Overlooked Issues in the Ethics of Care

Abstract Very often, but particularly in France, the free giving of blood is derived from the notion of *person*. Persons who need blood as a treatment would be respected by the free gift of other persons which take it for a duty. As honorable as this deduction may be, it seems too be false and it would be possible to derive from the notion of *person* a right to retribute the giver for his/her « donated » blood. But this argument supporting the payment of the collected blood, if it denounces a mistake, does not necessarily means that it becomes a maxim for action. I can give my blood, be convinced of the right to be paid for a blood collect and refuse to be paid for my gift. This is a proof that it would be other sources, more obscure and less rational, of the notion of *gift*. We attempt, in this chapter, to detect a few of them while searching how they are compatible with a rational approach of ethics.

Keywords Free donation of blood · Imaginary · Mythology of blood · Payment of the collected blood to the giver · Person · Rationalizing feelings · Religion and ethics

Blood transfusion only gives rise to debates in France when there are scandals, and those debates are explosive, as we saw in the 1980s and 1990s. Moreover, once some administrative and preventive precautions have been taken, it does not seem that it is being widely discussed anymore on an ethical level, as if the questions linked to transfusion were more fully settled than in any other part of the ethics of care. Discourses easily become axiomatic: it is well accepted that blood donation must be anonymous, that it must be free, that it is – or would be – contrary to ethics to sell one's blood or to know to whom one is giving it and for the recipient to know his giver. It is surprising that these axioms seem obvious, and even more surprising that those who, in ethics, go beyond the level of the general opinion organize fewer symposia on questions related to blood transfusion than on other issues of medical ethics,[1] which appear nonetheless quite closely related, such as organ transplants. Could not blood be assimilated to some tissue? There is another clue which draws

[1]The symposium organized by the Académie Éthique Médecine et Politiques Publiques on 8th February 2018 (www.IAMEPH.org) is a fortunate exception [32].

J.-P. Cléro, *Reflections on Medical Ethics*, Philosophy and Medicine 138,
https://doi.org/10.1007/978-3-030-65233-3_4

one's attention: when one looks for publications on transfusion in the catalogue of the Bibliothèque Nationale de France, one finds essentially French ones, with very few foreign documents, even Anglo-Saxon ones, and almost all those publications share the same orthodoxy about anonymity and free donation, as if these were intangible dogmas, and as if any position that moved away from those dogmas could not have the same ethical value, or even as if it purely and simply contravened ethics.

We would like to revisit some reasons invoked to support these dogmas or this supposed knowledge – even if unsure whether those reasons have ever been explicitly given – to see if they support argumentation, in other words to test whether the relative ethical silence surrounding issues of blood donation is linked to a refusal to challenge prevailing orthodoxy or a fear of arguing, and to query whether beliefs have not replaced argumentation, which nonetheless usually characterizes ethics as compared to morals and religious positions. We are obviously paying that new visit and undertaking that test to ask if an argumentation could not reassert itself in those sectors from which it seems more easily excluded than from others. Our goal is not necessarily to ruin the 'dogmatic' character that is displayed in some fields of ethics, and especially in this one. It is quite possible that, in some domains of ethics, it is difficult to avoid attitudes that are satisfied with *rationalizing* or *formalizing* feelings or affects which are not entirely rational. It could be that, despite our preference for a rational ethics, for an ethics of discussion, it is not really possible to establish such an ethics in all the domains in which issues of care arise, and that some of those domains at least can only be conceptualized by way of myths, beliefs and metaphors which have reached us from ancient times, and of which it is difficult to have a precise memory. It is not impossible that, in many cases, our attitudes towards issues linked to blood are only founded on those archaic ways of feeling, thinking and acting, even though the aim of ethics, put in place and developed for a few decades, has been to rationalize them as much as possible. One must become aware of those archaisms and perhaps better understand our attitudes in order to question them.

4.1 Blood Is not a Medicine as Others

Even though it is made part of our everyday life, drained of affective significance, and in a way made universal by the care and specific treatment it must go through to be transfusable, blood is not a liquid like others, and there remains something about word of what Poe says in *The Narrative of Arthur Gordon Pym*: '*blood* [of] that word of all words – so rife with mystery, and suffering, and terror', of these 'vague syllables' that fall 'chillily and heavily' 'into the innermost recesses of [our] soul'[2] – even though it is ostensibly a question of rational discussion of medical ethics.[3] It is not only its content or its materiality which make it singular. It is also due to its situation at the crossroads of acts for which we intend to sketch the beginning of an explanation.

Let us start nonetheless with its content. Even when it is at the heart of the medical act of transfusion, blood is not the same as a medicine. In a medicine, a molecule has been developed to solve a problem. The solution that the medicine supplies to the illness has gone through the intellectual and technical process of the making of that molecule. Admittedly, all the effects that a medicine can produce on an organism are not guaranteed from the beginning, but the cause, which determines those effects, is quite well known in most of its aspects. Blood, when transfused, even though it is especially well known and treated in a way that only blood-products are administered, contains so many more components that its entirety could not be dominated in an intellectual synthesis, as is the case in most medicines. Or, if the blood for transfusion can be assimilated to a sort of medicine, it is because what is true of those medicines – that we do not always know exactly how they produce their effects – is also true of it. Its singularity and complexity however, do not only come from the many ingredients it is composed of, but also – and this remark leads us to an altogether different order of considerations – from the fact that the blood that is received by a patient has already irrigated another body. Its very particular characteristics come from there – from being a component in another body before becoming part of mine own when it was transfused into me. Admittedly, that component is not exactly like an organ that is being grafted. It is not given in the same condition as those of a part of a liver, a kidney, a heart, a cornea, but it is clear that discussion of transfused blood cannot but be led in this direction, which is not the case with most medicines.

Thus – and this is what distinguishes it from an ordinary medicine – blood is a direct link with the other, even though one tries to attenuate it, or conceal it behind the modifications it goes through to be assimilable. There is something kept 'of the other' in this strange thing that is detached and administered to the patient. It keeps something of 'the other' under the form of 'thing'. Similarly, despite what those who work in the CECOS (Centres d'Étude et de Conservation des Oeufs et du Sperme Humain <Centres for the Study and Preservation of Eggs and Sperm>) say, there is an irreducible 'remainder' of sexuality in the medical acts of ART <Assisted Reproductive Technology; PMA in French: Medically Assisted Procreation> and surrogacy, there is also something 'of the other' in blood pouches, an irreducible remainder of otherness, even though it is asepticized and unrecognizable because of

[2] The end of the quotation from Poe is, as Bachelard translates it in *L'Eau et les rêves*: 'Comme cette syllabe vague – blood – détachée de la série des mots précédents (. . .) tombait, pesante et glacée, (. . .) dans les régions les plus intimes de mon âme!' ([1], p. 84). One can read that excerpt in *Aventures de Gordon Pym* published in the Pléiade volume which contains Poe's *Oeuvres en prose* [29]. *The Narrative of Arthur Gordon Pym of Nantucket*, Chap. III, in Poe Edgar Allan, *Poetry and Tales, The Library of America, Literary Classics of the United States*, New York,1984, p. 1035.

[3] One could say the same of blood in French, of 'dam' in Hebrew.

the preparation work they undergo. As, from far away, archaically, any link – to men or to God – is a blood tie, blood, even that of transfusion, forms an alliance and a social connection in a way that no other product can do. Even though the power of the direct or secondary effects of some medicines are frightening, none creates the same fear – even though it were felt only symbolically – nor the same reverence as those closely associated with transfusion, a fear and reverence which can only be felt in a lesser degree in relation to other medicines, even one that is very costly and used only as a port of last resort.

It is difficult to see blood, even when it is kept in pouches, as a simple thing, which could be deprived of all its enhancements, which are actually more negative than positive,[4] even though the happiness of recovering can be associated with them. The curative features of blood are ineluctably linked to the imaginary characteristics attributed to it, which are similar to those displayed by Diderot in a thought experiment in his famous *Letter on the Blind.* For a sight which seems unbearable to look at as soon as blood is included in it is perfectly bearable to listen to. Diderot's blindman is, in that respect, more insensitive than the sighted man.[5] He is indifferent towards the blood economy as any ethics of sighted people will weave it, or, if he cannot escape creating an ethic of blood, it is quite different from the economies made by those who are quite likely to feel moved by the sight of blood.[6]

Thus, though it can be detached from the human and the living being from which it is being drawn, while nonetheless still belonging in some way to their lives, blood cannot be understood as a thing without being at the same time endowed with symbolic functions. Blood speaks a language which only it is able to speak. What it says is only possible through that signifying reality which is its own. 'Speaking with one's blood', 'writing with one's blood' are common expressions used by very different authors, even those committed to imparting meaning through concepts rather than images.[7] We resort to invoking it – not without some pomposity that is liable quickly to become irritating – in the moments when rational evidence may be most lacking.[8] This is why evidence by blood can

[4]As Bachelard says, when he sketches the poetics of blood after Poe, he does not challenge the existence of that poetics, the happiness of which he nonetheless denies: 'C'est une poétique du drame et de la douleur, car le sang n'est jamais heureux' ([1], p. 84).

[5]'Quelle différence y a-t-il, pour un aveugle, entre un homme qui urine et un homme qui, sans se plaindre, verse son sang ?' <What difference is there to a blind man between a man making water and a man bleeding in silence?> [15].

[6]One could similarly show that imaginary feature of behaviours towards blood, as 'one needs all of Moses' ingenuity to imagine that blood could be completely evacuated from the flesh where it was circulating, as Bentham underlines in relation to kosher meat' ([2], p. 169).

[7]Is it not expressed in Nietzsche? 'Reading and Writing' in *Thus Spoke Zarathustra* begins thus: 'Of all what is written, I love only what a person hath written with his blood. Write with blood, and thou wilt find that blood is spirit' [25].

[8]It will be difficult to say that a mathematician writes his demonstrations with his blood, even though he must die in a duel, like Évariste Galois, on the morning following one of his most beautiful discoveries, or like J. Cavaillès, who was shot for his resistance activities against the Nazis, after writing *On Logic and the Theory of Science* [4] while in prison.

validate a testimony,[9] which is always a delicate moment of rationality. The unfailingly visual and affective experience of that singular liquid, the memory of that experience, and the designation of that experience and memory very easily constitute the signifying of the discourse that we are making or that speaks to us, but that signifying, as such, is obscure. Although the discourse to which it gives rise to is slightly less obscure, and often allows one to discern whether one is speaking of the thing or of the metaphor, it remains potentially contradictory.

Life and death are present in what is called *blood*. That which gives life[10] is also some refuse of life, like the blood of wounds and periods. However what bleeds also triggers some empathy, a sort of compassion, at the same time as it creates disgust: one feels that it is somewhere between life and death. Blood both tears apart the pure and the impure, degradation, ruin and salvation, and runs them together. It is in turn that which sullies and that which elevates and validates. Kosher meat, which some of us consider to be purified, is only made so at the price of a terrible haemorrhage. Blood can be the hateful blood of crime and revenge [*Rev*., VI, 9–10],[11] but it is also a sort of symbolic vaccine, which washes men, and can be what unifies them [*Eph*., II, 13].[12] Thus Paul says 'that flesh and blood cannot inherit the kingdom of God; neither doth corruption inherit incorruption' [1 *Cor*., 15, 50], but he also says in *Romans*, that if that the blood is Christ's, it justifies [*Rom*., V, 9].[13] The spilling of blood can be the sign of a curse, but it can be also that of salvation and purification, as in 1 *John*, 1, 7.[14] It is a leitmotiv in discourses on the New Testament. It is at the very heart of the Eucharist, when one must drink the blood: the 'repugnant' – who would wish to drink a cup of blood? And who would think of blood when drinking wine? – takes on a highly symbolical value, for that is how one imitates Christ and identifies with him.[15] Nothing is closer to the repugnant than the sacred which takes hold of it to give it another meaning. It seems that the sacred cannot operate without being close to what repels, and without mixing with it to exalt it. The spatiality of those crossings

[9]Pascal says it perfectly in fragment 663 of Lafuma's classification: 'I only believe histories whose witnesses are ready to be put to death' [26]. Paradoxically, one would not risk being killed for what one cannot demonstrate or prove.

[10]'The life of the flesh is in the blood', as is written in *Leviticus*, XVII, 11.

[11]'And when he had opened the fifth seal, I saw under the altar the souls of them that were slain for the word of God, and for the testimony which they held; and they cried with a loud voice, saying, How long, O Lord, holy and true, dost thou not judge and avenge our blood on them that dwell on the earth?' [3]

[12]'But now in Christ Jesus ye who sometimes were far off are made nigh by the blood of Christ' [3].

[13]'Then, being now justified by his blood, we shall be saved from wrath through him'. [It is of course God's wrath].

[14]'The blood of Jesus Christ his Son [the Son of God] cleanest us from all sin'.

[15]'53. Then Jesus said unto them, Verily, verily, I say unto you, Except ye eat the flesh of the Son of man, and drink his blood, ye have no life in you. 54. Whoso eateth my flesh and drinketh my blood, hath eternal life; and I will raise him at the last day. 55. For my flesh kismet indeed, and my blood is drink indeed. 56. He that eateth my flesh and drinketh my blood, dwelleth in me, and I in him' ([3] *John*, VI, 53–56).

should be much more refined, for blood ceaselessly links the interior to the exterior. Bleeding dangerously carries outside what is intimate, but it can also flood the body if it is drunk or inoculated, at the risk of disgusting or alienating one, making one a foreigner to oneself. That contradictory jumble of affects and feelings can easily be referred to in all sorts of acts, such as giving and receiving, which change its direction and meaning.

All our examples come from the Judeo-Christian tradition to illustrate the metaphorical and symbolic nerve centre that blood is, but that does not mean it is the only tradition to have shown it. The Greeks felt the same visual fascination for blood, which they understood as the occasion of a fall[16] and of a sacrificial atonement. In numerous cultures, blood has the ambiguous and contradictory role of creating segregation, as when, for example, one gets married within the same social class or caste, but it also serves to create links that transgress the limits of those castes or classes, thereby merging them.

4.2 The Images of Blood Are Deeply Rooted in the Collective Consciousness

II. What we have called a rather impenetrable sentimental and affective jumble nonetheless follows some logic, even if not a conceptual one. Conversely, even though blood is the means of transfusion, made within the framework of care, that is, in conditions which are mainly rational and meant to be so, its treatment does not follow the same conceptual logic as most care. As we have said, one does not receive blood in the same manner as one receives some antibiotics, or some antiviral, even though the drip conditions might be the same. Transfused blood is shared blood. This aspect must now be studied more in detail from an ethical point of view. Transfusion is not a simple ritual of reintegration into the world of those in good health from which one has been temporarily excluded. By giving me his blood, someone has collaborated to my reintegration into that community, not only by

---

[16]As in the well-known episode in B. IV of *The Republic* in which Socrates pictures Leontius, Aglaion's son, who, 'coming up from Piraeus along the foot of the northern wall on the outside and [noticing] some corpses lying beside the executioner, felt the desire to look at them at one moment and turned away in disgust at the next. For a time he struggled and covered his face; then, overcome by his desire he opened his eyes wide and run toward the corpses. « Look for yourself, you wretches, » he shouted, « and fill yourselves with an image of the beautiful. »' (439e-440a). (Plato, [28], p. 421).

symbols and speech – even though they permeate the whole operation – but in reality, by directly sharing their strength – even though it were in a manner mediated in a particular manner.[17] The donor's way of being united with the other – one could almost say with 'some other' – bypasses and transcends the person, by digging into something deeper than it.

In transfusion, the concept of the person is not necessarily central. In a typical blood donation, donor and recipient are usually unknown to one another: the recipient, if they wanted to say thank you for the donation, would be unable to thank anybody in particular, and that may be better, for that thanking might be indefinite and endless. If one said that anonymity preserves people from a relation that would be too heavy or would make them suffer in the case of ungratefulness, we would answer that though it is true – and we will come back to this issue – it was indeed not established for that reason, and that such an ideology of the person only conceals a reality that was already established long ago – that of creating a stock of the blood which moved the donor and the recipient apart. The feelings, through and in which we live such anonymity are not only those of a disguise of the person either. They are also positive. Though anonymity protects, as we will see, against the unlimited character and the potential excesses of donation, one must also fear a sort of arbitrariness which we feel less, and perhaps mistakenly so, when we are administered medicines. Anonymity, which fills the recipient with fear, is experienced in these questions: Is the donor healthy? Am I not going, because of this donation, to contract an illness that they might have, without knowing it, or without revealing it? Am I not going to be relieved from one illness at the price of contracting another, which is not known for the moment but which is as dangerous or even more dangerous?

We could – as we have already started to do – put next to each of the acts and feelings that we have pictured, sentences from the Bible, from the Old Testament, or even from the New, which follows a similar logic towards blood. There is a particularly striking analogy between the regulation – which, as though spontaneously, arises from the difficulties encountered in transfusion – and some biblical prohibitions. Some prohibitions which have been imposed over the past decades by public authorities after very serious transfusion accidents, or more generally, contamination, are quite similar to the prohibitions the Ancient Hebrew Bible. Curiously enough, the rules which, usually, in modern law, are not written in the imperative, and which in tone and mode are far removed from biblical prohibitions, become much more abrupt when the transfused are to be protected against donors who are suspected of being contaminated, or when citizens must be protected against murderers or criminals.[18] The unqualified exclusion contained in the rules that govern transfusions has often been noted by commentators: thus no-one who has received a

[17]The place and moment of the giving blood and of transfusion have been separated for a long time.

[18]J-L. Gardies, in *L'erreur de Hume*, legitimately insisted in this regard that modern laws differ from archaic ones in making minimal use of the imperative mood, and in the relative absence of the idea of prohibition [18].

transfusion can give blood anymore, following a logic of contamination which does not take into consideration any degree of probability. In 1997, people who had resided in the British Isles for more than one year all in all (between 1980 and 1996) were excluded from donation because they might be unwitting transmitting agents of bovine spongiform encephalopathy. I think it is quite clear now that the logic which governs blood transfusion is not fundamentally a logic of persons, if only because it chooses to protect one group against another, which is no less made up of people. In reality, blood gives rise to ethical rules the logic of which is far removed from a regulation of *persons* – as in the logic of Kant's ethics – and is quite persistent, despite its archaism and the fact that our ethics have been moving, for centuries, towards more abstract concepts like that of *person*. That move, which Ricoeur has depicted in *The Symbolism of Evil*,[19] makes ethics evolve from culpability and scruple, through the *pure* and the *impure*, the *stain* and the *sin*, to the concept of the *person*, and is never radical enough to preserve itself, even in its most recent manifestations, from the most archaic references. The true driving force of the *person* often lies in the ethical categories that those who support its conceptualization think they have overcome.[20] In reality, the ethics of the person only *seem* to be more rational than others. They only *appear* to monopolize rationality. Being at once contradictory and radically under-determined, they nonetheless continue to be animated by forces that have nothing to do with reason itself.[21]

[19]Which is the first part of B. II of *La philosophie de la volonté* [30].

[20]Admittedly, blood is entirely absent from Kant's morals. However, blood can still reappear, for example in his endorsement of a justice that kills, that spills blood, through the use of the death penalty. The logic of capital punishment does not arise, however, from the idea of person, since the death penalty might be prohibited in the name of that notion at least as easily as it is admitted by the strange Kantian argument that the person of the man sentenced to death is respected by that punishment. The notion of *person* is, as usual, equivocal, and accommodates directly opposing attitudes. This means that, in reality, it is always other considerations, some of which lack any obviously rational ground, that determine what one believes to be entailed by the notion of the person.

[21]We have already demonstrated this in relation to several texts. *Personne et anonymat. Du mauvais usage de la notion de personne*, [7]; 'Réflexions critiques sur l'usage de la notion de personne en éthique médicale' [8]; 'La solidarité peut-elle se substituer à la valeur d'autonomie ?' [9]; 'Personne et altérité dans l'utilitarisme' [10]; 'Qu'est-ce que soigner ?'; 'Le soin est-il une aide ?'; 'Qu'est-ce que l'autonomie ?' [13]; 'Has the Care in Psychiatry Other Characteristics than those it has in the Other Fields of Medicine? [11]; 'Y a-t-il, chez Stuart Mill, une spécificité de l'éthique entre les morales et le droit ?' [12]; 'Une pensée de l'existence à l'épreuve de l'éthique des soins. Les contradictions de l'éthique médicale' [14].

4.3 The Logic of Acts Involving Blood – Be They Murder, Revenge, Punishment, Sacrifice, or Transfusion – Is not Unrelated to that of Feelings and Passions

Passions do not operate like concepts, and they cannot be linked, one to another, as reason would link concepts. However, it is possible to understand how they work, in a way that is not without similarity to what happens in physics, when one understands, thanks to reason, the operating rules of beings or events which are not thinking beings or events of thought – with the difference nonetheless that passions, which are not events of reason, are nevertheless events of intelligence or thought.[22]

Hume explained in the second book of *A Treatise of Human Nature*, which considers passions, how what we call a 'logic of passions' ([22] 1991) worked, by showing how in the psychic system each of them is associated with others following an order which has its own necessity and proceeds through diffusion, transfusion, contagion, contamination, transfer, and identification – an order which is more or less perfect, but obviously always imaginary. He showed moreover how authority and sympathy were the two poles of passional phenomena, the associative rule of which he delineated.[23] The reader will recognize, in the words we have used, terms which all preserve a meaning in discussions about blood circulation itself and the affects that are especially attached to it – particular sympathy towards one whose blood is flows out, or who sheds their blood or accepts that it is shed, fear of being contaminated through blood, fear of real or imaginary dangers when it is transferred from one individual to another, and the game of the forces which are communicated in one direction rather than another, or which balance one another. Now is not the time for writing in detail about the passional operation of feelings linked to blood transfusion. We are going to consider only a part of the passional religious discourse, for we think that it underlies the logic of blood collection and donation in a particular way. It is only by making use of multiple features of the religious discourse and feelings, rather than of a more rational ethical discourse organized around the person, that we can in fact think about the issues linked to blood exchange.

Before establishing it, we need to introduce one last analogy, which, as we will see, will lead us quite a long way in posing some ethical problems, among which is the issue of free donation. We are referring here to the analogy that seems to exist between blood and money, or cash. In *Elements of Physiology*, which is about the physiological imagination which is our focus here, Diderot, for whom the function creates the organ,[24] and who, consequently believed in some versatility of organs when some necessary function must be fulfilled and is not fulfilled by the appropriate organ, suggested that each organ played its part in the body. In such

[22]This is what Descartes meant when he called them 'passions of the soul.'

[23]We have dedicated several works to that association of the passions and its rules [5, 6, 22].

[24]'Tous [les organes] ont leur vie particulière. [...] Si l'organe vit, il a donc une vie propre et séparée du reste du système' <Each organ has its own life. [...] If the organ is alive, then it possesses its own life separate from the rest of the system (https://books.google.fr>books) [16].

conditions, he could very well have seen blood as a sort of universal ferryman, transmitter, intercessor, which would ensure the necessary communication between all the organs, in its function of tissue, which reminds us of the role of undertaker of crossings that we noted at the beginning of this chapter. We underline the fact that it is exactly the role that classical economists give to money in economics. Though money is a good, it is what allows the other goods to be brought into a relation with each other, to be counted in the denominations of the others, and to be exchanged for each other. Economists – Marx first among them – often adopt metaphors of blood and its circulation to speak of money and its own circulation.[25] They meet another analogous problem situated comparatively at the same level. There is some affinity between the issue of usury and that of free donation of blood: that people can make money from their own blood could seem as scandalous in another period of time – and still does today – as making money with money. It was a long time before some economists – usually in utilitarian circles – lifted the prohibition that weighed for a long time on usury under the combined effect of Aristotelianism and Christianity. It was for a long time regarded as an equally unnatural perversion of exchange to make money with money, since it was not supposed to relate to itself, or to measure and exchange itself for itself, as it is an unnatural corruption to exchange one's blood for money. Just as one does not have to pay for the means of making payments, one should not have to pay for a vital solidarity which is at the very foundation of our humanity either. To make people pay for one's blood is in a way as sacrilegious as using money to distort the true meaning of work and goods. To understand this, and maybe to start moving away from this type of argument, which may be, as we will see, at the foundation of the sacralization of the body and its transformation into a sanctuary by the law, which leads to prohibitions directly opposed to *habeas corpus*, one needs to look at things from a little distance and understand the affinity between discourses on blood and religious discourses from which we will never escape if, as Hume says ([22], 1956, 1971), religions constitute the necessary chain of connection and co-ordination between the passions. What we must now consider is how unsurprising it is that the discourse on blood remains a religious discourse, and to understand the way it which it must be so because, conversely, religious discourses are always, at least in our countries and cultures, discourses about blood, or at least involving blood. If one were surprised that economic considerations on blood are in one sense hampered by religious considerations, one should, conversely, underline that religious discourses are fundamentally economic discourses. Thus, by combining the two approaches, we would like to show that if it is very difficult for our ethical discourses to get free from religious discourses, that is because a blood

[25]'En tant que valeurs, toutes les marchandises ne sont que des mesures déterminées de temps de travail coagulé' <all commodities are merely definite quantities of congealed labour-time> [24]. Here Marx re-used a sentence he had already used in *A Contribution to the Critique of Political Economy*.

economy is at work in those religious discourses.[26] The discourse of blood and the economic discourse reciprocally serve one another, and each acts in turn act as basis and reality for the fictions of the other.

4.4 The Religious Discourses We Know in Our Culture Are Discourses of Blood and About Blood

A great number of the books of the Old and New Testaments can be interpreted as continuous variations on the theme of blood. Blood and the acts involving it are indeed always presented as regulated by a sort of economy which has its own rules which, even though their principle is not in reason, and consequently always escapes it at least partially, can nonetheless be understood by reason, and are formalizing – and let themselves be formalized – up to a certain point. The logic of punishment, that of retribution, and that of honour, all rely, at some point or other, on blood as a sort of 'master signifier', and involve a logic of the sacred which is always a logic of exchange, which troubles reason even though it can find explanations and even rationales for it, which are not its own, and treat them in a sort of axiomatic way, by accepting that it cannot understand them completely. For the fact that reason cannot understand a value like blood does not prevent it from using it as a signifier to inform a certain number of discourses. Anyway, are there values that reason cannot completely understand and in relation to which it can do more than develop an incremental understanding of how they operate? 'J'ai versé telles gouttes de sang pour toi' <I shed these drops of blood for you> has no directly rational meaning, but this strange clause that Pascal puts in the mouth of Christ[27] may have another meaning in other ways and according to different postulates, according to some thought of the other. One must, to understand it, understand the idea of a certain market which has its own scales. What is always striking in expressions of the sacred is their incredible utilitarianism, even if it may be an imaginary one, for one cannot really disentangle what is real from what is fictitious. 'The church of God, [has been] bought with his own blood', as is written in *Acts*, XX, 28. Christ's blood allows for a redemption (*Eph*., I, 7).[28] It is used as a bartering system in *Rev*., V, 9.[29] With all due respect to those who wanted to 'personalize' Christianity, when it comes to blood Christianity only rarely locates us, at least in the hands of those who wrote the

[26]With a little more space, we could have developed a curious analogy between the circulation of blood and that of language, to the advantage of the word *plasma* which, in Greek, means *fiction*.

[27]The text of 'The Mystery of Jésus' also includes the following fragment: 'Veux-tu qu'il me coûte toujours du sang de mon humanité, sans que tu donnes des larmes?' <« Do you want it always to cost me the blood of my humanity while you do not even shed a tear?»> [26].

[28]'In him we have redemption through his blood, the forgiveness of sins, in accordance with the riches of God's grace'.

[29]'You are worthy to take the scroll and to open its seals, because you were slain, and with your blood you purchased for God persons from every tribe and language and people and nation [...]'.

Gospels, in an ethics of the person. The religions that we know refer to a much more archaic ethics. Saint Paul's comparisons of sacrificial value which allow for a fortiori arguments – if the sacrifice of animals makes it possible to purify some mistakes or impurities, how could the sacrifice of who has never sinned not have the properties which would allow for redemptions of a very superior value?[30] – are a long way removed from the theme of the person.

A certain number of thinkers of our time are still affected by what could be called 'arguments' of that order. It even seems that such speculations and the weighing of purity and impurity could, apparently and unfortunately, be reinforced by some phenomena that, over the past decades, have marked the history of transfusion, which has been conceptualized in very dangerous and mythical terms, and in a way which seemed to be giving a real foundation to simple phantasms. This is how one can argue – quite wrongly, actually– by relying on some supposed reality, that the therapeutic value of the blood that is given is higher than that which is sold, that the blood of some categories of men is more impure and may sow death more than that of other categories. In this way, the blood of some foreigners can be viewed – quite wrongly, but with all the awkward endorsement, though it might have been involuntary, of public authorities – as more dangerous than the blood of autochthons, and, among those autochthons, that of homosexuals as more dangerous than that of heterosexuals. Thus the necessary prudence which consists in excluding some individuals from blood donation, may have seemed to endorse, and directly and awkwardly to legitimize racist, xenophobic and homophobic discourses by offering them, in addition, unhoped for agreement and a semblance of rationality. Quite unfortunately, these divisions coincided with those drawn in tales of which the far right has always been, and still is, the champion. Archaic divisions, which make prostitution, blood and foreigners coincide are coming back, from the Bible, in these terrible discourses.[31] That which was a phantasm of infection by blood in *Revelation* [4] [XVI, 3, 4, 6] may even have seemed to take on the appearance of reality and to be confirmed by things themselves. And to crown it all, some illnesses can be contracted by blood as well as sperm, which is even more linked to the most hackneyed of superstitions: was it not believed in the Middle Ages that sperm came directly from blood? Because of an illness like AIDS, sperm and blood had their destiny once more linked in the 1970s and 1980s. Blood opportunely offered to

[30]Saint Paul who shows very well that Christ's sacrifice is a surpassing of animal and human sacrifices ([3] *Heb.* IX, 12-14; X, 4) – since for a long time the first born in some families were sacrificed, and a bargain with God was needed to substitute animals to the first born that God seemed to be asking from human beings. Reason revolts at this idea and recoils from the sort of analogy that Saint Paul dares write, which at the same time rationalizes and offends reason. When blood is spilled to pay for the fault of the one whose blood is spilled, it is the logic of retribution and punishment: hurting somebody who has inflicted hurt seems to belong to a certain order, but when blood is spilled and the one whose blood is spilled is innocent, it must be something utterly different, and that blood must be worth something completely different, e.g. that it redeems the sins of the other men. But what seems to be rational is not at all so. Saint Paul is rationalizing something which is actually absurd and aberrant.

[31]As is seen in *Acts*, XV, 20, 29; XXI, 25 [3].

the populist slogans the appearance of reality they needed to reduce to a common denominator behaviours which, however, had nothing in common in reality: those of homosexuals, prostitutes, drug-addicts and immigrants. The history of transfusion sometimes contributed to give 'good' reasons to tie more closely together what opinion and intuition already 'instinctively' put together, with the appearance of science on its side. In this game of reversal of the for and against, as Pascal would have called it, the utmost unreasonable and irrational seemed to have coalesced with science and prudent behaviours, which made it possible to cloak ostracism and rejection of the other with a varnish of respectability.

4.5 Is It the Concept of Person that Makes the Selling of One's Blood Scandalous?

It is clear that the debates around the free donation of blood, that some would like to close down even before they have been opened, belong, at least in part, to that mixed discourse on religion and trade. What makes the selling of one's blood scandalous if not the fact that the value of blood is sacred for us?

Indeed, the arguments usually used to justify free donation that are claimed to be based on the person are not really convincing, and it is but too easy to find others which plead the reverse, that is, are in favour of the payment of the person whose blood has been collected. Here, as elsewhere, the concept of *person* gives rise to all sorts of contradictions, and consequently does not allow us to solve ethical problems, or, if it seems that it does, it does so only rhetorically, by covertly relying on concepts other than that of the person itself.

A person would not respect themselves by selling a part of their body, even though it would regenerate itself.[32] There is an opposite way of looking at things, which may legitimately be supported by changing one's point of view and supporting counter-arguments: why is a person more respected by not being paid for a favour they do for an individual or for the community, than they are by being paid for it? It is possible to give a scandalous turn to the argument for free donation: is it ethical to refuse to compensate the person who does an important favour to others, or even more, an irreplaceable one? Admittedly, one understands that respect for the person of the recipient entails that they never know their donor, while the respect for the donor entails that they never know the recipient, who, if they knew the donor, would have contracted a potentially infinite debt to them. However, the use of Kant's morals to attempt to solve the issue of free blood donation is quite fallacious and unproductive. Obviously we are not denying that Kant made a distinction between the moral value, the sentimental one, and the price by which goods are alienable. But it should also be recognized that, among the duties that Kant

[32]Jurists sometimes talk of the *non-patrimonial* and *non-commercial* dimension of the human body, when, in other times, one would have talked of its *inalienable* nature.

foregrounded, he did not forget, among the examples chosen in *Groundwork of the Metaphysics of Morals*, to include the scrupulous respect for commercial contracts, prominently among which feature the selling and buying of goods on the market?[33] We cannot see why it would not be ethical to pay a fair price for a service. For us, the real problem lies more in the setting of a fair price than in any opposition to the strict non-payment of the price itself, whatever it may be. Otherwise, one must believe that there is something in blood that makes it unsellable, whatever its price.

It might be argued that blood cannot be sold because the biological production of blood is not, properly speaking, for a body, a work like any other, as the technical and chemical preservation of that blood until its presentation in usable pouches indisputably is? But do we only sell work, work coagulated into goods, to use the words of Marx [24] in an expression which is quite interesting for our purpose?

What is it that cannot be 'paid' in blood? There are favours that I pay for, and which are paid to me following very different moral and sentimental modes. When I am granted care with the utmost moral attention and the most exquisite delicateness, I still pay for it. However embarrassed I feel when I have to pay the fees, I would not dream of not paying the doctor who was considerate and treated me with such exquisite delicacy, precisely because he is considerate and courteous, nor of deducting the value of consideration and courtesy from the price of the consultation, to the point of making it nothing, if not negative, which I could apparently legitimately do if the values of consideration and delicacy had no price, or by being beyond price compromised the setting of any price.[34] Such an attitude, made general beyond what strictly concerns free blood donation, would lead to a radical reversal of values, even though it is true that one is always embarrassed to offer payment for a favour that is performed to a standard beyond expectation. But have we not already adopted a reversing attitude when we do not pay the person who 'gives' their blood? Admittedly, there are apparently excellent reasons not to pay them. Everything happens as if there existed no possibility of 'paying well' and as if donation, being good in itself, did not require to be paid, which is absolutely wrong, and seems very reminiscent of Plato's bad faith towards Sophists, who were discredited by the words he gives Socrates for having had their teaching services remunerated. Must the services of solidarity – provided we consider the obligation to give one's blood as an obligation – be as free as that of truth for Socrates? However, one could ask Socrates why truth would lose any value by being imparted in the course of paid

[33]Admittedly, the trader who sells at the same price to the shrewd buyer and to the naive child does not necessarily do so out of duty, even if their action is compatible with their duty, but it is a basic duty to respect contracts.

[34]This sort of paradox is often found in morals when it moves apart from common opinion. Schiller could criticize Kant in his parody of the strange scruples of the conscience of morals. Here, we have a great 'scruple of the conscience'. 'Scruples of Conscience: Gladly I serve my friends, but alas, I do it with inclination and thus I am frequently nagged by my lack of virtue. Decision. There is no other advice, thou must seek to despise them, and do with disgust what thy duty commands' ([31], vol. I, pp. 299-300); trans. J. Timmermann, in *Kant's Groundwork of the Metaphysics of Morals*, ([23], 2007, p. 152).

lessons, or in paid articles in books and journals? Why should it only ever be 'given'? It is when we pursue such a question that we come to suspect that the resolution not to pay the right price for blood collection is made under cover of a moral disguise: for there are fair and unfair payments. If the resolution were not never to pay but to find the fair price, would the market not eventually set the 'fair price' for all the products on it? Even though it were granted a sacred value, blood is not 'priceless', economically speaking.

And, above all, in a dangerous binary logic of all or nothing, of the brutal reversal of the for and the against,[35] is it because 'some things' are priceless that they can and should be freely given, because they are not priced? One often finds the sign of a taboo, in ethics and morals, in a rule that applies absolutely and without nuance, even though the situation would allow for a position which gives rise to the grading of degrees of probability. And is it true because the sacred is priceless? Even though food, housing and education were sacred, in the sense that one could not deprive a man of them without disrespecting his humanity, are not victuals, houses and schools priced? If health is priceless, as is often quite imprudently repeated, why should what directly produces it, such as nutrition, housing and education be priced? There is a prejudice according to which a donation must be good, which is not true. Even though writers who talk about it tend to find it has a moral value, it can nonetheless erect social values on disputable and ill-defined foundations which contrast with the clear and determinate boundaries characteristic of money relations. I feel I owe something to the gift-giver, but how am I to know (and to demonstrate to their satisfaction) that that debt has been discharged? Though it is true that, in a certain number of acts of selling and buying, money, whether one has a lot or a little, puts pressure on the seller and buyer and structures their relations, how can one believe for a moment that donation does not do something similar? Donation can sanction, ratify, establish more or less durably – sometimes for life – an inequality between the donor and the recipient. If I pay for something, whether I am poor or rich – provided I am not completely destitute – I can detach myself from the seller from whom I have bought something; I can less easily detach myself, at least without being accused of ungratefulness, from someone who has given me something and to whom I cannot give it back[36].

Apart from the apparent contradiction of its actually bearing a cost price, donation may be ambiguous in other senses. There is, for instance, no such thing as a pure donation, and in this failure in its purity the gift makes a complete mockery of the independence it assumes on the part of giver and receiver by undermining it on both sides. Either donation is imposed on those who receive it, and do not possess the

[35]Several examples of which we have already noted in the previous chapters 2 and 3: see p. 28, p. 50, above.

[36]Lacan wrote in the Book Two of the *Seminar*, precisely in the Chapter XVI of *La Lettre volée* *<The purloined letter>* : « Chacun sait que l'argent ne sert pas simplement à acheter des objets, mais que les prix qui, dans notre civilisation, sont calculés au plus juste, ont pour fonction d'amortir quelque chose d'infiniment plus dangereux que de payer de la monnaie, qui est de devoir quelque chose à quelqu'un ».

possibility of not receiving it, or it is imposed on those who give it: does not the law, by demanding that it be free, impose donation, which is not without its contradictions? Anonymity is, in this sense, a protection against the bottomless debt that donation would otherwise create.

Thus without wanting blood collection to be anything else other than a donation, since, after all the crises it has gone through and despite them, the system has proven its worth, and there seem to be enough people who support it for it not to require radical change in this regard, I only wished to take some distance from justifications which rely on the concept of the *person* or on a moral law of a 'personalist' type;[37] for, though it is true that one can draw arguments in favour of the principle of anonymity,[38] it is not true that one can draw from it some duty of gratuitousness. For the argument of the 'personal' foundation of free blood donation to be convincing, one must not only think that one has a duty towards oneself not to sell a part of one's body, but only to give it, but, in addition, one must think that the duties that one believes are incumbent on oneself must always be the foundation for the duties one believes are incumbent on others. In other words, one must assume that, when an individual does not conform with the duties they are supposed to have towards themselves, one can compensate for their failure, even though one would not have to suffer from it, not being them, and not depending on them. Personalistic morals is possible only by relying on indemonstrable and improbable axioms which may define a morality, but never an ethics. One always comes back to the obviousness of a taboo on the issue of blood. There is a taboo about blood as there is one about sex. Venality is readily invoked in order not to pay for the blood given in a transfusion, as been done in condemnation of surrogacy. One might also happily rely on the scorn cast on the latter to discredit the idea of payment for giving blood. The analogy however, does not provide much clarity, and does not take things further, for it supposes that whoever is won over by the argument thinks that payment for carrying to term the foetus which in usual circumstances is the product of a sexual act between consenting people is a bad thing in itself, whereas it is no more or less disputable than paying for blood donation. Behind the moral argument of free donation, to which, as if by magic, are attributed all sorts of virtues, and which would be to the glory of the donor possessed of healthy (if not quite holy) and safe blood, and would be the only one to be perfectly generous as they are perfectly disinterested, one must nonetheless measure what each gives up to the State, which not only refuses to pay – which might satisfy the taxpayer – but also prohibits individuals from making payments. What about, then, in these conditions, *habeas corpus*? Does the State not behave as if, without any answer being possible, it

[37]As is the case for Kant: 'So act that you use humanity, whether in your own person or in the person of any other, always at the same time as an end, never merely as a means'. (*Groundwork of the Metaphysics of Morals*, sect. II, in ([23], 1996, p. 80).

[38]Note though that anonymity does not offer only advantages: there are circumstances in which one might wish one knew the name of the woman or man who has donated her or his blood, especially in the case of a contamination by a virus which was not known at the time when the transfusion was made.

possessed peoples' bodies with which no one is permitted to deal with as they want, even though the body in question were one's own? This ode to gratuitousness is not as innocent as it seems, not only because it is quite indecent that it should be whoever refuses to pay that sings it, but also because the State then acts as if it were the absolute master of what I can and must do with my body, as if it knew it better than anyone, and in any case better than myself. It can defend my body against the use I make of it, even though I am not causing any harm to anybody. Why should those who want to give their blood be prevented from doing so in exchange for money? One knows the generous arguments which aim to defend the poor against themselves by depriving them of rights to which, like all men, they are entitled. Would there not exist any way of preventing abuses – and we do not ignore the manipulative and complex bends those abuses might follow[39] – rather than ban the act of giving one's blood in exchange for money by outlawing it? Like the heart-throb of the beautiful soul for the welfare of mankind that Hegel criticized ([20], 1939, I, p. 309 ; 2018, p. 217), the beating of the heart for poverty always seems a little suspect when it emanates from people who not want to pay. Is not their citizenship – and in some way, the person of the poor that one pities and defends against themselves – denied when they are deprived not only of a little money or of a way of earning some, but thereby of the right to decide between possible means of accessing some money, simply because one thinks that one knows better than them what they must do and not do?

One should, of course, not be naive and ignore too easily that, behind the poor, who sell their blood, lurk the multinational companies which buy it and get richer thanks to a re-sale that the poor could neither undertake nor organize on their own. Such an international traffic should obviously not be allowed to proliferate, but means to prevent it might be found without forbidding any payment to whoever gives their blood. We know that migrants sell their blood and organs to raise money to pay the people-smugglers who will get them to Europe. But even though the law cannot remain abstract and cut off from social reality, it cannot abolish all difference between the facts and the law, and declare that poverty should deprive the destitute of any use they want to make of their body. In that case, what is offered in exchange for that which takes the form of a prohibition on making use of oneself by oneself? If one responded bluntly that the State does such things all the time, the argument would be disastrous, for it would reduce the idea of *habeas corpus* to little more than a dead letter. Moreover, should one not be careful not to pretend hypocritically that States are hermetically sealed, so that the law established in each of them applies strictly, without the law of the other States interfering with it, limiting it, overcoming it or preventing its application? 'In France, 70% of immunoglobulins are coming

[39]François Pilet, an economic journalist in Switzerland, a member of the International Consortium of Investigative Journalists, who was awarded the 2013 Jean Dumur prize for his book 'Krach Machine', explained a little time ago – on the 8th February – at the symposium of Paris-Descartes University on *La Vente des Produits Sanguins, Précarité et Vulnérabilité*, how drug addicts, for example, use their dealer's donor card to continue to give their own blood and get the money they need to buy drugs.

from American paid blood donations. If tomorrow President Trump closed the export of American plasma, there would be no available clotting factors in Europe.'[40]

Free donation – provided one is at least a little honest in defending it – has its reasons that the heart does not know. In any case, it derives from something quite different, from the idea, which seems much more rational, perhaps too rational, of the person. Those 'reasons', since one must call them such, are on the side of the righteous who, by spilling their own blood, could redeem the sins of the others,[41] which is understandable only in terms of quite a different logic. Reason understands rather that it is for everybody to make sure they do not commit any sin. 'Loving is giving one's blood'[42] has no more sense in that logic than Saint Paul's logic in the *a fortiori* arguments that we quoted. However, despite all the irrationality of the ethics of blood, which we have noticed in the Old and New Testaments, it nonetheless enjoys a certain hegemony over the ethics of the person, which latter are too contradictory to provide a solid foundation for the former by endowing some of its aspects with comprehensible and plausible meanings. Nonetheless, whoever receives the blood of another in a transfusion can hardly avoid unease at the thought – an unease only enhanced by the conditions of the administration of the blood products themselves – that another has spilled their blood, as they themselves could have spilled their own blood for another. Transfusion, all things considered – for the donor does not die for the recipient! – even through the sanitary and commercial filters that it interposes between the donation and the final reception by a patient, puts one on the path towards understanding something that is of the order of a graft, something that intelligence should not reject under the pretext that it cannot fully understand its logic.

[40]Michel Monsellier, the President of the Fédération Française du don du sang bénévole, said this during the debate we were mentioning previously: see the previous note.

[41]«Now the God of peace, that brought again from the dead our Lord Jesus, that great shepherd of the sheep, through the blood of the everlasting covenant» ([3], *Heb.* XIII, 20). Just before, Paul inflamed Christians with these words: « XII. 3. For consider him that endured such contradiction of sinners against himself, lest ye be wearied and faint in your minds, 4. Ye have not yet resisted unto blood, striving against sin. » The just must be ready to die for sins to be redeemed. The image of Christ as a lamb is also in [3] 1 *Pet.*, 1, 19: « Forasmuch as ye know that ye were not redeemed with corruptible things, as silver and gold, from your vain conversation received by tradition from your fathers; But with the precious blood of Christ, as of a lamb without blemish and without spot ».

[42][3] *Revelations*, I, 5.

4.6 Conclusions

We would like to conclude this chapter with seven reflections, of which the sixth opens up a generalization of what we have discovered that will be pursued in the next chapter.

1. First, we do not think that the world of goods is radically opposed to that of donation. There is no reason why money, because of some fetishism of commodities, should possess some inherent characteristics which are destructive of morality, co-operation, solidarity and social links. Commodification and gratuitousness are not necessarily irreconcilable. There is always the possibility, in a complex enough economy, not only for a duality of the pure and antithetical forms of donation versus the market, but for a reality full of parallel economies where the donation and market spheres intersect in manifold ways, and manifold different hybrid configurations exist. Blood collection is typical of this type of crossing. The authors who reflect on blood donation from the economic point of view, like Frey or Zelizer [33], are legitimately looking for some sort of intermediary path between the logic of the State, that of the market, and that of donation, which are not necessarily in an antinomic relation.
2. This first way of looking for intermediary paths which allow for blood to be collected by organizations which are, up to a point, independent from the State, is not unrelated to the two logics we have constantly encountered and set in opposition, at least at a certain level, but with a view to preparing the ground for some relative conciliation at another. It is clear that the ethics of the person is ill-equipped to conceptualize the issue of blood donation exclusively in its own terms. It is equally clear that the ethics of the person aspires to be – and thinks it is – more rational than the archaic exchanges that used to exist in the religions we know, but it is absolutely powerless, despite the best efforts of those who want to derive ethical positions from it, because those positions appear at best inconsistent and at worst flatly contradictory of each other. Consequently, the only consistent ethical position in that domain which deals with all those reversals from the pros to the cons, tends to be a rationalization which goes against the fundamentally irrational positions which have all the characteristics of taboos. It is not impossible that an irreducible part of ethics cannot make do without Hebrew, Greek and Christian myths, to which apparent rationalizations of the *person* give a more respectable appearance in the eyes of people like us, who have a certain idea of reason, though it alone cannot constitute the driving force and power of ethics. The great power – we are not talking of truth here, for that power can arise from foundations that are completely wrong – of mythical resources compared to a concept like that of *person* – which from the very first seems more rational and which is envisaged, as Ricoeur showed, as an overcoming of the ethics of the pure and impure, of the ethics of scruple – comes from their narrative quality: they tell stories in which we can include elements of our own lives and of the lives of other men, which makes them more credible. This is also, of course, what makes them more dangerous.

3. One might wonder whether the insistence on the *person*, the weakness of which we have tried to demonstrate, could be linked to the fact that transfusion is, at least symbolically speaking, a direct threat to the limits of the individuality of the persons. If I can receive the blood of another, or the tissues of another, is my body not something that could be completely invaded, flooded by the other, in its very intimacy? In that case, the only way to resist that disagreeable feeling is to claim a 'personal' foundation for that which nonetheless blurs all the limits, biologically and in reality. The person is truly a frail mask.
4. I gave my blood in the past, and it never occurred to me to be paid for it, so why this apparent paean to commercialization? This is precisely the place to highlight the distinction between a personal moral option and an ethical rule, and above all, to distinguish clearly between the two and keep them separate. The result of trying to make a personal decision universally results in bad ethics. Universalizability is a necessary condition for a decision to be moral, but universalizable reasons can be advanced without one thinking for a moment that they should effectively be made universal. An ethical rule has a real application. It leaves the greatest space to a decision of the other rather than to my own, or to the decision I would have taken in the stead of another. If one should try to universalize the maxim of the action that the ethical rule recommends, it is more in the sense of knowing, first, if that maxim is compatible with the decisions which others would have taken in similar circumstances, and which, on this very same topic, might differ from it, and, second, if it limits my existence in any way other than imaginary ones.
5. Lastly, it has not been my intention to upset the reader. My position is critical in the most ordinary way. Just as Kant did not claim the privilege of telling the mathematician what to do in mathematics, or any other scientist how to practise his science, just as he viewed the morality he developed not as his invention, but simply the outcome of reflection on what he took to be most common morality – a claim which involves difficulties all of its own – I do not wish to argue that 'free' donation should be replaced by a commercial system of blood collection. Being rather sentimentally in favour of free donation, according to a *prima facie* position – as R.M. Hare called it [19] – I simply wanted to study the arguments which are used to establish such and such prescription, proscription, rule, prohibition or position. This has led me neither to set aside common appreciations, nor to adopt sceptical positions, which are actually untenable in practical issues, but to challenge the supposed arguments which in fact conceal intuitions which might be correct, even if founded in and supported by false reasons.
6. We have to return to the mythical aspect which is supposed to combine and integrate with more rational elements. It shares a symbolic character with what Hare called the critical level [19], for one must be wary of what seems to be mythical in our eyes, and seems, in a way, to be purely intuitive. The intuitive may in fact be an amalgamation of the intuitive and the symbolic, which we can no longer distinguish, and the strata and levels of this amalgamation can fold into each other, at points we can no longer locate. It seems that there are two pitfalls that must be avoided in order to understand the organization and relation of the

mythical and the critical. The *first* is failing to multiply the levels of moral thinking sufficiently, and only distinguishing, like Hare, two levels, which are notoriously not enough.[43] The *second* is, on the contrary, considering the play of the folding of the levels one by another in too naïve or superficial a fashion, as is sometimes done in mathematics when escalating levels of symbolism permit the solution of problems that the intuitive level makes it possible to pose, but not to resolve.[44] The play of folding does not seem to occur in ethics as it does in mathematics: a surpassing in mathematics reduces to almost nothing the level that was at its origin. In ethics on the contrary, there is that game of folding by ever-higher levels of the symbolic, but the levels that are felt to be intuitive are most of the time retained in some form. There is no final containment of the intuitive by the symbolic, with no possibility of coming back to the former, as is the case in mathematics, in which today's intuitive is no more than yesterday's superseded symbolic which, having overcome a previous intuitive, has had to be overcome in its turn.

7. Lastly, in order to have a closer look at the last point, we would simply like to note that what we have called 'one of the most forgotten issues of ethics of care' – transfused blood – shares this status not only with some fields of medicine, but also with the way pharmacy is structured at a national and international level, from the manufacture of its products to its distribution, by way of financing and systems of reimbursement, when they exist, as in France. It is clear that there is an urgent need for a monograph on this point which would identify the issues and investigate the entire context in which the essential elements which enable the patient to recover their health operate, including of course the care of the doctor, but also far removed from either him or individual consultations. There is nothing fortuitous in the silence that is maintained on such an important point, and were it once broken, it is likely that most of the ethical problems related to care would appear in a different light, as the ethics of psychiatry clearly illuminate the problems of risk in all the other fields of medicine.

[43]Even though Hare had the merit of launching the movement, he could hardly have avoided treating it in too simple a way, by missing its complexity.

[44]To solve his problems linked to the cycloid, Pascal had to compose spaces that did not correspond to the intuition he had of them. Even though his heart felt that space had three dimensions, the needs for the demonstration made him built four-dimensional spaces, for which he apologized, but which were nonetheless necessary to build the centres of gravity of volumes generated from cycloids. Hilbert, for his part, surpassed Desargues' theorem by showing that his solution was valid only in very particular spaces and not universally, as its creator could only have imagined in the seventeenth century [21].

Bibliography

1. Bachelard, G. (1973). *L'Eau et les Rêves. Essai sur l'imagination de la matière*. Paris: Corti.
2. Bentham, J. (1823) *Not Paul, but Jesus*, under the fictitious name of Gamaliel Smith, Esq. London: Hunt.
3. The Bible. The Acts; The Epistle of Paul the Apostle to the Hebrews; The Epistle of Paul the Apostle to the Ephesians; The first Epistle General of Peter; the Revelation of Saint John the Divine, in: The New Testament, Authorized King James Version, The Gedeons International.
4. Cavaillès, J. (1960). *Sur la logique et la théorie de la science*. Paris: PUF.
5. Cléro, J. P. (1985). *La philosophie des passions chez Hume*. Paris: Klincksieck.
6. Cléro, J. P. (1995). *Hume. Une philosophie des contradictions*. Paris: Vrin.
7. Cléro, J.-P. (2001). Personne et anonymat. Du mauvais usage de la notion de personne. In *Les cahiers du Comité consultatif national d'éthique pour les sciences de la vie et de la santé, n° 27, avril* (Vol. 2001, pp. 35–38).
8. Cléro, J. P. (2011). Réflexions critiques sur l'usage de la notion de personne en éthique médicale. In *Deux siècles d'utilitarisme*, under the direction of M. Bozzo-Rey and of É. Dardenne (pp. 211–231). Presses Universitaires de Rennes.
9. Cléro, J. P. (2013). *La solidarité peut-elle se substituer à la valeur d'autonomie?* Rouen: PURH.
10. Cléro, J. P. (2015). Personne et altérité dans l'utilitarisme. *Ethics, Medicine and Public Health, 2015*(janvier-mars), 82–90. Sffem: Elsevier Masson https://doi.org/10.1016/j.jemep.2014.09.001.
11. Cléro, J. P. (2016, December). Has the Care in Psychiatry Other Characteristics than those it has in the Other Fields of Medicine? *Revista Româneasca ventru Educatie Multidimentionala, 8*(2), 45–56. https://doi.org/10.18662/rrem/2016.0802.04.
12. Cléro, J. P. (2016). Y a-t-il, chez Stuart Mill, une spécificité de l'éthique entre les morales et le droit? In *Philosophical Enquiries, revue des philosophies anglophones*, December 2016, n° 7.
13. Cléro, J. P. (2017). Qu'est-ce que soigner?; Le soin est-il une aide?; Qu'est-ce que l'autonomie? In *Le soin, l'aide, care, cure*. Rouen: PURH, in collaboration with Annie Hourcade.
14. Cléro, J. P. (2017). Une pensée de l'existence à l'épreuve de l'éthique des soins. Les contradictions de l'éthique médicale, in revue Cités, Collectif *Jankélévitch*, n° 70, juin 2017.
15. Diderot, D. (2000). *Lettre sur les aveugles à l'usage de ceux qui voient* (p. 38). Paris: GF Flammarion.
16. Diderot D. (1916). The letter on the blind. In *Diderot's early philosophical works* (trans. M. Jourdain). The Open Court Publishing Cy.
17. Diderot, D. (2004). *Éléments de physiologie* (p. 335). Paris: Honoré Champion.
18. Galois, É. (1962). *Écrits et mémoires mathématiques* (éd. R. Bourgne et J.-P. Azra). Paris: Gauthier-Villars.
19. Gardies, J. L. (1987). *L'erreur de Hume*. Paris: PUF.
20. Hare, R. M. (1981). *Moral thinking. Its levels, method and point*. Oxford: Clarendon Press.
21. Hegel, F. W. (1939). *La Phénoménologie de l'Esprit* (trad. J. Hyppolite). Paris: Aubier.
22. Hegel, F. W. (2018). *The phenomenology of the spirit* (trans. T. Pinkard). Cambridge: Cambridge University Press.
23. Hilbert, D. (1971). *Les fondements de la géométrie*. Paris: Dunod, 1930, *Grundlagen der Geometrie* (ed. Teubner). Leibzig.
24. (1971). *L'Histoire Naturelle de la Religion, et autres essais sur la religion*. Paris: Vrin, 1956, *The natural history of religion* (ed. H. E. Root). London.
25. Hume, D. (1991). *Traité de la nature humaine de Hume*. Paris: GF-Flammarion. See the Preface to B. II on Les passions.
26. Kant, I. (2007). Groundwork of the metaphysics of morals. In I. Kant (Ed.), *Practical philosophy*. Cambridge: Cambridge University Press.
27. Marx K. (1993). *Le Capital, Critique de l'économie politique*, L. Ier (4th ed., p. 45), Paris: PUF.

28. Nietzsche, F. (2013). *Thus spoke Zarathustra. A book for all and none* (ed. by A. del Caro and R.B. Pippin, trans. A. del Caro). Cambridge: Cambridge University Press.
29. Pascal, B. (2004). *Les Pensées*, Br. 553, Sel. 749, 751, in *Les Provinciales. Pensées*, La Pochothèque, Le Livre de Poche/Classiques Garnier, Paris, p. 1317. 1995, *Pensées*. London: Penguin Books, p. 290.
30. Pilet F. et Lelièvre F. (2013). *Krach machine*. Paris: Calmann-Lévy.
31. Plato. (2013). *Republic*, Books 1–5 (ed. & trans. C. Emlyn-Jones & W. Preddy). Cambridge, MA/London: Harvard University Press.
32. Poe, E. A. (1951). *Oeuvres en prose*, traduits par Baudelaire, NRF La Pléiade, p. 531. 1984, *Poetry and tales*. New York: The Library of America, Literary Classics of the United States.
33. Ricoeur, P. (1988). *La philosophie de la volonté*. Paris: Aubier (Vol. I: *Le volontaire et l'involontaire*, Vol. II: *Finitude et culpabilité*).
34. Schiller F., Werke, L. C. Hanser (1987) In 5 vols, Darmstadt, vol. I.
35. Symposium of Paris-Descartes University on La Vente des Produits Sanguins, Précarité et Vulnérabilité, Feb. 8, 2018.
36. Zelizer, V. (2019). *The social meaning of market*. Princeton University: Princeton.

Chapter 5
Imaginary, Myth and Concept in Medical Ethics

Abstract So there is no possible ethics without mythical and imaginary components; as soon as we want to rationalize ethics it seems that we cannot help but leaving remains aside. But there must be an agreement on what is called *mythical aspects* of ethics. Sometimes, the matter is to assign to ethics tasks that are impossible to realize; for instance when one promotes a medical follow-up fo every patient through out their existence. Another figure of mythic or imaginary could be the awkwardness in the presentation of treatments or in the choice of a treatment that can promote the imaginary idea that a treatment is better than another, without the least objective reason. Lastly, there is an ultimate meaning of imaginary of which the positivity is more admissible: decisions -those of the patient as those other doctor-are always taken with an uncertainty. The Bayesian calculus, that we prefer among the calculations in decision models, does not compel the actors to one position rather than another; it only points with which chance to be right we make or have made this choice rather than another in given circumstances. The decisions have been staged by the ancient Greek theater whose myths are constitutive elements of ethics in conjunction with the rationalization by the calculus we have spoken of.

Keywords Aeschylus · Ambiguity of the notion of *myth* · Bayesian approach of probabilities · Concept · Diodorus Cronus · Imaginary · Mathematics and myths · Myth · Palliative care · Pandora · Randomization

5.1 Introduction

Our medicine is obviously no longer that of the Ancient Greeks. We can cure illnesses that in their eyes were incurable, our doctors have vastly reduced the suffering which results from illnesses as well that which results from their treatment. When they are not curable, which still often happens, our illnesses have become less disabling and may not prevent us from working, or even from living a fully engaged life until the end, admittedly with a few drawbacks, thanks to efficient treatment, for

J.-P. Cléro, *Reflections on Medical Ethics*, Philosophy and Medicine 138,
https://doi.org/10.1007/978-3-030-65233-3_5

months or even years. Varied and powerful imaging guides the surgeon who, thanks to the superior preparation of his operation, is less surprised by what he discovers when he must act and only hurts the body in as minimal a way as possible. It can guide the radiotherapy which now benefits from means to target the destruction of the damaged tissues more precisely. Medicines are powerful. Their secondary effects are better known and defined, and better care can be taken to minimize them. They therefore do not compromise the quality of the life of the patient so much, and though they may be still be ill, they quickly seem to recover good health, and can continue in apparent good health for a long time. All these features, which everyone can flesh out with their own experience, and which could be multiplied many times over, are moving us away from Greek medicine. However, despite their indefinite, constantly increasing quantity, they cannot make us forget the ancient Greek origins of our own medicine. Its main patterns and axes were first established more than two and a half millennia ago and are still often form the basis of our medicine, despite the increasing multiplication of differences to which we have alluded, which constantly move us further away from their origin.

To be convinced, one need only witness the particular way caregivers listen at a conference on the ethics of care, to contributions on ancient Greek medicine, and the singularly animated debates that follow, to feel the caregivers' impression that such medicine and its presuppositions lie at the very root of what they do and are capable of generating the reasons which validate the aspects of their task they hold dearest.

It might be argued that these are merely listening attitudes, which have neither more nor less meaning and influence on behaviour than the attitudes we adopt when we listen to a tale. They link us back to the attitudes we used to have towards stories when we were children, but they tell us little about the deep structures of real acts of care and the ethical modalities which penetrate, permeate and accompany those acts, and form and reform the cognitive structures we use to organize them. Admittedly, we would not go as far to say that there are not significant differences between the Greek ethics of care and ours, but we cannot but be surprised by the deep affinities that will be revealed as this chapter unfolds. The Greeks had their myths, and we have our own: since the seventeenth century, we have believed in progress, as if a unique teleology supported the fragments of our knowledge and techniques, reduced them to a common denominator, and carried them, in a broad movement, towards increasingly impressive successes. The Greeks did not generally share our belief in progress, even though they had a glimpse of that idea, and they certainly did not have the same conception of history.[1] However, we remain strongly attached to some of the dimensions that they identified in the medical act: how could one not be affected,

[1]To generalize, they thought that the Ancients were worth more than us and that they knew what was true, because they lived closer to the gods.

when reading Hippocrates' oath, by the idea that whatever happens to the patient, and all the more so if things go wrong for them, the patient will be able to count on the doctor who will not abandon them? Not being abandoned by your doctor, even though your life seems to have lost any social value and has become a burden for you, being able to count on your doctor may well be the central value of the ideology of care, which was reborn in our society a few decades ago.

Moreover, it is not only in its deepest convictions that our ethics of care resembles its Greek counterpart: it also resembles it in echoing the manner in which the Greeks made the criticism of ethical concepts an important element of ethics itself. Sextus Empiricus, a doctor of the second and third centuries AD, wrote a 60-page text entitled *Against the Ethicists* [21], in which he did not recommend that anyone, and in particular not the doctor, should avoid holding moral convictions or act as a godless and lawless debaucher, but in which he also demonstrated, in page after page, that the difficulties of ethics, which are still noticeable today, present real pitfalls into we might slip in widely varying circumstances. We are often incapable of agreeing on a classification of values because we cannot convincingly argue, and make everybody agree, with our conviction that it is such or such value rather than another one which, in given circumstances, must govern our actions. We may say that ethics consists in vanquishing, in any given situation, the multiplicity of morals, but we do not know for sure how to render that victory legitimate. We do not know the object of ethics. We cannot raise it to the height of an indisputable science, nor even to that of an art, and we are consequently incapable of teaching it. However, from this position that may seem desperate, Sextus Empiricus [21] formulates a crucial value of ethics, which is the opposite of dogmatism. Far from being a simple exercise in the promotion of values – which morals, religions or any other belief could be – ethics cannot do without a criticism of values. That which it supports always retains something which is derived from that criticism, without completely undermining itself for all that. At a time when some religions try to capture ethics to their profit, transforming it into a sort of branch of themselves, it is worth recalling the iconoclastic dimension of ethics, by remembering that it lies often more in the 'suspension of the judgment' than in the reckless promotion of a certain number of values that believers want to impose on patients. Sextus saw how the promotion of supposedly positive values quickly turned into a machine for making men despair [21].[2]

It is that triple, mixed and fluctuating relation of myth or of the mythical element with conceptual reason that we want to examine here – myth sometimes makes the conceptual function ideological and fallacious, but myth also offers, at other times, an auxiliary function, with reason exercising, but only up to a certain point, a critical function towards the myths to which we ascribe different meanings.

[2]We will come back to this point at the end of the text.

5.2 First Version of the Mythical Function: A Strange Expression in a Recent Report on Ethics

A report on ethics written eight years ago (it was written at the beginning of President François Holland's term of office and at his request) by the French Comité Consultatif National d'Éthique (national committee on ethics),[3] deals with the end of life, takes stock of the present situation in France, comparing it to what happens in other countries, and recommends a few changes in the medical organization in that field. This is not the place for us to judge all the aspects of the form and matter of that report in its entirety, but let us focus on one of the assertions it contains, which is a leitmotif: palliative care fills the last hours, days or weeks at most of a patient whom doctors have given up on curing after trying in vain. The failure of the active, interventionist, Promethean medicine permits it, but only very late on, perhaps much too late, to pass control to a medicine which nurses while assuaging and comforting, without any hope of curing, and which, in a strange swing of the pendulum, and in almost shame-faced remorse, displays the Epimethean values of feeling, softness, and passiveness rather than obsessive activism.

Quite reasonably, the author(s)[4] ponder this sudden rupture in the goals of care which may give the impression of a desertion of the patient who has reached the end of their life. They criticize the way in which *palliative* care has become coincident with end-of-life medicine, though it should never cease to be continuous care for the patient, maybe with peaks of activity or hyperactivity at crucial moments of the illness which may require heavy surgery which entails particular monitoring, or radiotherapy, the secondary effects of which are not trivial. The authors then embark upon a sort of apology for the high value of *continuity* and inveigh against the anti-value of *discontinuity*, when care must be implemented late or even ultimately and is legitimately felt by those to whom it is administered as corresponding to their entering the anteroom of death. This would not be the case if its values accompanied the care journey of the patient, and not only during the terminal phase. This way of criticizing discontinuity then takes on the appearance of a promotion of the continuity of care, about which one might wonder whether it is much better than the discontinuity that should be abolished.

To put it clearly: what is reprehensible here is not the trial of that discontinuity of care which may be problematic when the patient must change doctors, departments, or even hospitals, from an intensive episode of care to the cessation of curative care itself, but the exaggerated inversion of that discontinuity. Should doctors continuously examine the patient: is that psychologically or economically reasonable? Is it likely that everybody lives for the continuity of their health while other people would be at the service of that continuity? Must continuity itself become an absolute value,

[3]It was given by Professor D. Sicard to President Hollande on 18th December 2012, and was entitled Penser Solidairement la Fin de Vie < Thinking about the End of Life in Solidarity> [22].

[4]Pr Sicard was not the only contributor to the report and took care to be guided by a committee in drafting it.

being at the basis, like a *basso continuo*, of the necessary discontinuity of morbid episodes? Undoubtedly, the desire for continuity of care appears at the moment when its discontinuity underlines the inconveniences of acute episodes, but should it become a 'reality'? Must it cease to be a sort of coveted fiction? If it ceased to be a coveted fiction, if it were reinforced into a sort of social requirement that would be more or less internalized by the individual, would it not become a sort of tyranny of health? Because the painful and acute episodes of a long and serious illness appear to the patient like the unconnected links of a chain which should never have broken, must continuity be insisted upon to the point of becoming obsessive?[5]

Such a claim for the continuity of care, against its discontinuity lived as a tragedy and in despair belongs to a sort of imaginary and mythic thinking. It reverses what should have remained a fiction into a position. The taste for and promotion of what is continuous[6] against the inconvenience of what is discontinuous is a mythical value, and plays a role that is more fallacious than essential, since everybody knows that it is impossible to establish such a continuity, that it is not even advisable that that value be really established and that it is better for it to remain a fantasy of the moments when discontinuity weighs on us.

We will draw from this first example of the difference between the two medicines, the curative one and the supporting one, a use that may be called mythic, according to which continuity is better than discontinuity. The first type of myth that we have just explored, if we are permitted to call that first attitude a myth, which is in reality closer to an ideology than it is to what is usually understood to be a myth – that is, a tale on a topic which cannot be completely rationalized, consisting in a sort of imaginary or affective remainder – is still quite close to the ideas of truth and falsehood. Relying on critical arguments alone, we might be unable to decide whether care should be continuous or discontinuous, and we might discuss for a long time just what should be understood by continuous care, but one nevertheless has the impression that we might also succeed, and that our dilemma could be solved by appeal to reasons. We would have come quite close to a distinction that the Ancients made between 'true opinion' and 'false opinion'.

---

[5]It is difficult here not to recall the words of Socrates, who in *The Republic*, ([19] 406 c-e, 2013, p. 301), asserts with Asclepius that 'knowing that a function has been assigned to each and everyone of those who are well governed in the state which they are obliged to perform, and that no one has the time throughout their life to fall ill and be treated, something would be absurd among the working classes, but which we don't see among the rich and those who are apparently happy'. Moving on at once to economic considerations, Socrates shows that caring for oneself all the time is a pleasure for the rich, which is quite harmful to the city!

[6]It is especially noticeable in Aristotle, even in his *Physics* and *Metaphysics*. The συνεχές, the continuous, cannot be artificial in Aristotle. It is reserved for nature, as underlined in *Metaphysics* book Δ, 4: 'All things are said to grow which gain increase through something else by contact and organic unity (or adhesion, as in the case of embryos). Organic unity differs from contact; for in the latter, case there need be nothing except contact, but in both the thing which form an organic unity there is some one and the same thing which produces, instead of mere contact, a unity which is organic, continuous and quantitative (but not qualitative)' (1014 b 22–26) ([3], p. 221).

However, the mythical character can also take on very different appearances and slip unnoticed inside rational arguments in a way which gives them structure without mimicking reasons, without, that is, coming close to being correctly described as either true or false. Some rational arguments stand in need this sort of gate, which is absolutely not rational and does not mimic reason.

5.3 Appearance of a Second more Positive Mythical Function

There are indeed other myths the foundational function of which is more relevant, owing to what could be called the good management of hope, the famous ἑλπις, which is to be found, hesitating, on the neck of the bottle from where all the other virtues (or vices?)[7] have fled because of Pandora's imprudence, and which Plato [19] theorized so well in *The Laws* (644 c-d), by giving ἑλπις, beyond its dimension of hope, its component of fear, thereby transforming the idea of hope into a remarkable symbol of human actions or of the human condition as a whole:

> *The Athenian*. May we assume that each of us by himself is a single unit? *Clinias*. Yes. *The Athenian*. And that each possesses within himself two antagonistic and foolish counsellors, whom we call by the names pleasure (ἡδονὴν) and pain (λύπηυ)? *Clinias*. That is so. *The Athenian*. And that, besides these two, each man possesses opinions about the future, which go by the general name of «expectations» (ἑλπις); and of these, that which precedes pain bears the special name of «fear» (φόβος); and that which precedes pleasure the special name of « confidence » (θάῤῥος); and in addition to all these there is «calculation» *<λοισμό>*, pronouncing which of them is good, which bad; and «calculation», when it becomes the public decree of the State <γενόμενος δόγμα πόλεως>, is named «law» < νόμος>. *Clinias*. I have some difficulty in keeping pace with you: assume, however, that I do so, and proceed.'

What follows is a very clear reference to the myth of Pandora and to Hesiod.

This imagining of hesitation on the line that separates the inside and the outside, the known from the unknown, and its Platonic theorization which aims at broadening the idea, cannot but throw us into paradoxes which, though they were stated in the fourth century BC by Diodorus Cronus [11] and were staged by very ancient writers of tragedies, do not concern us less, or even do not dismay us less, for we are still prisoners of the nets of their alternatives, and cannot solve them. I will come back to the dilemma of Diodorus Cronus, but let us first read a contemporary tale.

At the juncture of care and research – a juncture that is quite unavoidable in medicine – lies what is called *randomization*. It consists in giving the patient who is being treated for an illness that is difficult to cure, the choice between treatment T1 (which is the usual, procedural treatment that is used, and involves taking risks that are known in the given case) and a new treatment T2, of which there is some hope

[7]One does not really know.

that it could be better than T1, but which is not certain yet, not even its probabilities.[8] However, for the experiment to be successful, and to proceed blindly, the patient must not be left to choose between T1 and T2, but the experimenter must be given complete initiative and will build a distribution of them upon random chance. Without that clause, one could not know whether T2's curative effects are really superior to T1's. Thus, the patient is asked to choose not between T1 and T2, but between T1 and (T1 or T2), the doctors keeping, thanks to the second branch of the alternative, the possibility to apply treatment T1 or T2. The alternative is between a T1 treatment, which is statistically speaking well-known, and another branch of the alternative, T1 or T2, which the patient, out of principle and consent, is not master, and which is statistically less known, since that knowledge is still to be conquered, the attempt to conquer it will be more or less risky.

Law books and medical books have widely popularized the case of the American couple who out of principle accepted this type of choice, and then demanded that they 'benefitted' from the T2 treatment after choosing the alternative within the alternative, which was a violation of their initial agreement, and which was refused by the doctor-researcher, who could then only, within the framework of that violation of the agreement, offer the T1 treatment.[9] For our purposes the result of the trial and the end of that story are beside the point. We need however, to examine the two branches of the antinomy that we regard as inevitable, and in which conflict is more disguised than resolved by the play of consents. The experimenter can legitimately think that the alternative was clear and that he should not be obliged to change its modalities at the last moment. It is not for the patient or for his spouse to distort the terms of the initial contract which offered the option not between T1 and T2, but the option between T1 and (T1 or T2). Consent having been given to the contract as it was presented, it is not possible unilaterally to change its terms and modify a research strategy into a care strategy by putting forward an imaginary interest based on a calculation which may be wrong. This distortion does not have the meaning or the scope ascribed to it by its supporters, for nobody knows – the researcher no more than the couple – what real value the T2 treatment has, since the randomization system is precisely established to test it. The researcher was quite right to question the foundation of the couple's case, and even their good faith, when they changed their minds, and accordingly to refuse to apply the T2 treatment and to revert to the T1 treatment as the only option, as if the contract had never been offered or entered into.

Moreover, there is an ethical foundation for the position of the researcher and the patients who enter into randomized trials in good faith, since when I accept a system

[8]I am making a distinction here between the certainty that an event will happen (or has happened) and the certainty of the probability that it will occur (or has occurred), which implies that that probability lies between two values – as is the case in Bayes' probabilities, which are subjective chances of being right in expecting that the probability of the happening of an event lies between any two degrees that can be named. [5]

[9]The case was first presented in *The New England Journal of Medicine* by Eugene Passamani [18]; and then a second time in *Journal of Clinical Oncology* by E.J. Emanuel & W.B. Patterson) [12].

of research, I accept both that others may receive benefits from the objectification to which I submit my person, and that I may in fact have consented to my own disadvantage, even though neither I nor anyone else could have known that at the time. I am supposed to forego no advantage in the operation by which I treat myself as an object of knowledge, and am respected as a person through the requirement making my initial consent to randomization a condition of the trial.

However, and here is the delicate issue which justifies – contrary to the thesis of the researcher and the compliant and understanding patient – the antithesis which was chosen by the American couple, and which gives their complaint a universal value: if the doctor-researcher, who declares they wish me well and whom I trust, gives me the choice between T1 and (T1 or T2), is it not because they suppose that the T2 treatment is superior to the T1 treatment? Am I not correct in thinking that, even though they concealed that belief out of principle? Otherwise, why would they have conducted the trial and asked me to participate? From that moment the suspicion or some plausible reason for believing that T2 is superior to T1 arises – even though that reason to believe is not an ἐπιστήμη knowledge, a positive science – is there not a breach of what the patient owes to their own person if they meekly accept the possibility of the T1 treatment, even though they might have, so they think, the certainty of the application of the T2 treatment, which seems to have, in the eyes of the specialist, a good chance of being superior?

Whether one likes it or not, randomization puts the patient in a difficult situation, for they cannot realize that they hurt themselves before starting one of the two treatments, and their consent cloaks that apperception, but cannot but feel it when, changing philosophies – if I may say so – almost as soon as one of the treatments has been started, although they do not know which one, since there necessarily remains a doubt in the event that the state of the patient deteriorates, that he may not have been given the better treatment in the end, and may have accepted, through his consent, the worse one. Against a backdrop of non-knowledge – for ἐπιστήμη (positive science) acknowledges its limits on this point, and I do not know, in the end, if submitting to treatment T1 and depriving myself of that knowledge, T2 would have been better, or if receiving treatment T2 and not knowing, T1, in the end, may have been better – the presentation of the randomization process is such that it goes against the ethical principle of what I owe to myself. This is serious, as it is not impossible that the randomization system, which is so important for research, always leads to these or similar difficulties.

Here is the boundary between two logics that confront one another without either possessing the key to unlock and defeat the other, for why should the reasons to believe give way to the objective knowledge (ἐπιστήμη), when the latter does not and cannot possess its highest possible degree of certainty at the moment when it asserts its rights?

A rational practice falls into a snare which is quite difficult to avoid rationally. Well built though it may be, the tree of probabilities nevertheless results in a situation which drives the patient beyond the limits of reason. It is not impossible for another 'good solution' to the problem we have just encountered to exist. Very well, you might say, but why then speak about the mythical function of ethics?

5.4 A Third Mythical Function Illustrated by Bayes Rule and Aeschylus' The Persians

We must indeed go further in our reflection to find the mythical nature of the basis of the calculations that the Greeks ignored, since those calculations depend on the idea of statistics – which only developed much later, when J. Bernoulli launched it at the end of the seventeenth century with his law of large numbers [7] – and on its skilled application – as demonstrated by Bayes in the second half of the eighteenth century with his famous rule presented in *An Essay towards solving a Problem in the Doctrine of Chances.*[10] Bayes was a pastor of a dissident church of the Church of England, and did not care directly about Ancient Greece, but I will nonetheless argue that the basis of Bayes rule may be understood by means of an interpretation of the problem of Diodorus Cronus [11], which is itself, in many respects, the elaboration of a scenario presented by Aeschylus in *The Persians* [1].

The Greeks have too often been said to be incapable of thinking in terms of probabilities – though they were subtle in geometry – for lack of the economic and social conditions in which that mode of thought became possible, and which first emerged in 17th–eighteenth century Europe.[11] Admittedly, it is easy to see that the calculation of expectation or probabilities only started, properly speaking – by which I mean giving results that were undisputable, or at least undisputed – in the middle of the seventeenth century, but the establishment of ἐλπις, the exact thinking of the probable, and the denunciation of the paralogism that is usually committed in relation to it, were enough to grasp the essence of Bayes rule and the prudent use that must be made of expectations. We are encouraged in this direction by Aristotle, who examined the probable in myths – in tragedy in particular – in terms of an opening onto the future and who studied, because of that very opening, its generality which placed it so high above history [2].[12]

One could but advise the doctors who have not done it yet to have a close look at Aeschylus' play and meditate on Diodorus Cronus' paradox [11],[13] which consists

[10]Bayes [5] poses his problem in the following terms: When one only knows little about an event or a series of events, for example whether it occurred n times and did not happen m times, n and m not being large numbers, what chance of being right does one allocate a value to a probability that one decides about that event or series of events? Bayes gave a general solution to that problem which was in a way the reverse of Bernoulli's [7], who asked, from the moment one has a very important number of events at one's disposal. How many of them must be taken into consideration to appreciate their probability with a margin of error that suits one.

[11]This was the meaning of the thesis E. Coumet presented in his memorable article, ([9], pp. 574–598).

[12]The decisive text of Aristotle, *Poetics*, is quoted at length in the conclusion to this Chapter: see p. 98 below.

[13]What about the truth or falsity of judgments concerning an event planned in the future, which has not happened yet, and may never happen? Should one systematically say that the judgment is wrong if it asserts the existence of an event which, as will appear, will not have happened? Can it retain, in these circumstances, some truth?

in wondering if present facts can determine the accuracy and thus rule in favour of or against a prevision of a battle (or of any other event) which will take place the following day (or on any future day), and thus what is the status and the value of truths which it seems will not be decided on the following day or on a later day. Is it possible to decide for or against facts? It is important in studying this paradox to try and avoid using the kind of statistics produced by the system of the *Evidence-Based Medicine* (EBM) in an inappropriate way. The operation of this system is 'Laplacian', if one may say so. Just as the organization of astronomical science that Laplace imagined at the end of the eighteenth and beginning of the nineteenth centuries was located within an international system for which the Laws of Newton served as a contract among the astronomers who admitted it – not because they believed it was true, but because it made it possible to give direction to the work done by each of them by means of reciprocal corrections and amendments[14] – so today the prevailing organization of medical science assimilates the case of each patient, with its singularities and indefinite complications, into a set of similar cases, which make it possible to see better what is done and can be done for patients, and to build classifications and define protocols which, if not the most appropriate individually, are at least the best designed statistically, based on the treatments which have been tried in a multiplicity of cases which are apparently similar. A curious process is undertaken by doctors who are qualified in the sense of having learnt what statistics and probabilities are all about. They often give in to the strong temptation, often because they want to reassure a patient about their health prospects, in discussing with them the value of a treatment or seeking to convince them of the merits of some intervention, to use the statistics obtained from the sum of singular cases by folding it, without sufficient caveats or precautions, onto the singular case of their patient. The theoretical tool, which is extremely precious thanks to the information that EBM gives the doctor and the researcher, is then used as an argument of comfort, hope, and reassurance, to facilitate a decision in favour of a treatment for which the consent of the perhaps reluctant patient needs to be obtained. It is possible to tell the patient that their case has every chance of being in the 99 'good' percentiles, or 9 'good' deciles, or the 3 'good' quartiles, and that it is very unlikely that it will turn out to be in the 'bad group'. What happens, however, if it is discovered, after an laboratory test, or MRI scan or any other diagnostic tool, that the case lies in the 'bad' quartile, decile or percentile, despite the reassuring statements of 'Doctor-So-Much-ForThe-Better'?

The main risk is that of the loss of confidence which, admittedly, affects not the value of the doctor or surgeon's technical skill in practice, which can be preserved despite that loss, but that of his speech. Obviously, we do not wish to promote the opposite error by approving instead the words of 'Doctor-Too-Bad', which would be as inappropriate as those of his counterpart. What is at stake, because of its

[14]These endless corrections are made from the observations of some and the elaboration, by the most theoretical among astronomers, of 'small equations' which correct the big one – established by Newton – without changing it radically.

formidable impact on the relations of trust between doctor and patient – which should not be forfeited without benefit and hope of repair, during the short moment of the reading of a presumed result which has been imprudently *predicted to be good* although it is bad – is the rhetorical use of statistics. Many doctors consider discussion of probabilities as imprudent. In *The Persians*, Xerxes, the Persian king, considered that his victory on the Greeks was probable, unlike the Greek war leaders who, not trusting what the augurs told them, went on to win the war which turned into a disaster for the Persians. He who believes what an augur says, or who takes at face value what some ambassador who, being a double agent, pretends he is betraying his camp and announces its tragic future,[15] makes the mistake of imagining that the future is written in the same way as the present and the past, that good augurs – as if they existed! – and gods –as if they existed elsewhere than in men's beliefs – can 'know' about future in a way which the majority of mortals do not 'know'. However, far from those beliefs, true wisdom consists in considering that nothing in the future is ever written, that things that are estimated and calculated about future are not already distributed as behind a curtain, that nobody – neither the common people nor the augur, be they good or bad, nor the gods – knows by ἐπιστήμη how things will develop. Probability does not exist as something that is as real as the present or the past, so that its reality is only temporarily veiled. Probability does not exist. That does not mean that in many circumstances we do not calculate what consequences our actions will have. However, as long as those actions still remain unperformed, nobody knows the consequences, for there is nothing to be known about them by ἐπιστήμη <by some knowledge established with certainty>, since that knowledge does not have any object.

We once again come across Cronus' paradox [11] in relation to the following point: the probabilities of the reasons for believing are different from the probabilities of things or events, and, strangely enough, are not retrospectively sanctioned by their advent either. Someone may, in given circumstances and making the best use of the information that is accessible at that moment, decide to do something, in accordance with their reasons for believing that the situation is such and that things would go best if a certain thing were done, and yet find themselves apparently belied by the facts which seem to contradict their calculations. Have they been wrong? Not if that person took the most rational decision they could take at the time they took it. One can be right in believing that something will happen which nonetheless does not happen. One can even, once they know how things have turned out, continue to believe that they were right to decide as they did, when they decided and acted, even though circumstances seem to have ruled against them. One can also have been wrong to take a decision, given the circumstances in which it was taken, even though

[15]For, as the messenger tells the queen, Atossa, 'a Hellene, from the Athenian host, came to [her] son Xerxes and told this tale: that, when the gloom of sable night should set in, the Hellenes would not hold their station, but, springing upon the rowing benches of their ships, would seek, some here, some there, to preserve their lives by stealthy flight. But Xerxes, on hearing this, not comprehending the wile of the Hellene nor yet that the gods grudged him success, straightway gave charge to all his captains to this effect' ([1], I, p. 141).

in the end it turns out to have been the 'good' or 'right' one, and has apparently been sanctioned by the facts.[16] One has simply been lucky.

In his famous dilemma, Diodorus Cronus [11] establishes that the assertion that a naval battle will take place tomorrow is not necessarily wrong, even though one will come to know, in the end, tomorrow having become today, that it has not happened, because probability, while not existing in things, has its own ontological consistency, and is about the reasons for believing or not which are grounded in the reality of things. Its fiction, even though reality has not 'decided' yet, if one might so express it, is not groundless. This conception of the probable – which is absolutely not some hidden 'in itself' that should be revealed as if it were behind a veil of ignorance that the future, becoming the present, will eventually lift – allows for some action that is not hindered. Xerxes' mistake was to believe that things were written, though nothing is, and the calculation of probabilities, though it is the way, for the future, to be written – and provided it had any sense in Antiquity – does not provide any knowledge of things, but only allows one to act, precisely while ignoring them.

In addition to the dimension of a tale that we have developed almost exclusively in what we have written up to this point and to which we will return again, there is another dimension which results in the myth drawing closer to mathematics, and which we will only mention in passing here. The dimension of the *symbolic* is the written, algebraic, non-lived side of truths, and is quite distinct from the *imaginary*, that is to say from the obvious, lived side of truths, or what might be called the understandable or conceivable, meaning by that a kind of presence of the idea that would give us the illusion of possessing it in some way. Myth is not more obvious and self-evident than mathematical symbols. Admittedly, there is a certain type of production of symbols in mathematics based on other symbols that allow the former to be deduced, whereas myth does not deduce its signs from other signs, and considers only successive relations between the signs that have a value for successive sequences. However, there is, in both the symbolic and in myth, the same indifference to the immediate 'conception', from which arises the parallel need to make a particular effort to grasp them.

The only point that can distinguish – not oppose – the symbols of myth from those of mathematics is related to the sort of folding they each produce. Folding can happen almost without remainders in mathematics. When Hilbert [15] undertakes a folding upon Desargues and Pascal's problem [10], what remains from the approach of those two classics is a meagre lemma in a system that has just absolutely deposed their majesty, without comment. Desargues' theorem is forever attached to a space that is no longer current, which is at least deposed in its generality or universality. Conversely, even though myths too are folded in different levels of the symbolic, there survives from them a core that directs the link between the different symbolical levels to a sort of commenting activity, whereas mathematics does not move forward by commenting upon itself – except when the mathematician becomes an historian

[16]So many politicians launch elections campaigns that seem so impossible to win that one would not bet a cent on them, but end up winning!

of mathematics. It is possible to do mathematics without doing the history of mathematics. It is not possible to do mythology without, in some way or another, well or badly, doing the history of myths. This does not mean that the mythical discourse is of a different essence from the symbolical levels that comment upon it. It simply relates the successive levels in a different way from mathematics, where there is more of a dissolution of what is at the core of a theorem when it is superseded. The amalgam of discourse which constitutes the myth is, it is true, of the same nature as the whirl of symbols that comment upon it, but it has its own distinctive attitude, that mathematics does not share, towards the positions which are superseded in its progress. Mathematical criticisms do not let the positions they dissolve re-amalgamate, whereas mythical criticisms result in an amalgam which retains more of what came before and establishes itself much more on the basis of it.

5.5 That the Mathematical Should Not Be Opposed to the Mythical

One can see here that, despite the differences which have just been pointed out, it would be a mistake to oppose the *mathematical* to the *mythical*. First, because mathematics, the rational nature of which nobody will deny – and this is the very reason why we are insisting upon it now – could not work without something it shares with myths: namely a tale. A mathematical problem that is to be solved is at one and the same time a tale and the promise of a tale. A solution that did not 'tell' a story would not demand the least attention and could not be read without pain. Those who do not like mathematics see nothing in it but unimaginative tautologies where others, those who are engaged with it and take pleasure in it, constantly tell tales: they cannot see that it brings into play values that are foreign to deduction, which are nonetheless necessary to that deduction. In *Théorème Vivant*, Cédric Villani makes his reader aware that the best of his research proceeds by relating some narrative: this is how the most improbable phenomena, provided they are not impossible, take on the appearance of the tales of miracles which sometimes have their equivalent in the Bible.[17]

[17]Even though Villani never mentions his fondness for, or indeed makes explicit reference to the Bible. Everyone knows the story of the paralytic who lies in wait for the stirring of the water, though no visible sign has triggered it. There is an explanation for that phenomenon, which, then, is no more than quasi-miraculous, in an equation that Villani takes care to translate into a vernacular language: 'Imagine that you are walking in a forest on a peaceful summer afternoon. You stop by a pond. Everything is calm, there is not a breath of wind. Suddenly, the surface of the pond starts convulsing, everything is agitated in a great whirlpool. And then, one minute later, everything is calm again. [. . .] The Scheffer-Shnirelman paradox, [which is] certainly the most surprising result of the whole fluid mechanics [upon which Euler's equation rests], proves that such a monstrosity is possible, at least in the world of mathematics' ([24], p. 98). John's words about the water miraculously troubled by an angel before the impotent man in the pool of Bet-ghes'da thus receive ([8] *John*, 5, 2–6), two millennia later, some particular endorsement from the mathematician, even though not for the reasons he might have foreseen.

As to the probabilities discussed in this chapter, the narrative is so fundamental that it imposes frameworks upon the mathematician who, if he were to ignore them, could not find a solution to the problem he poses.[18] Just as a painter chooses an easel of a specific type of wood, a more or less tightly-stretched canvas and one made of one thread rather than another, and pigments of particular physical and chemical qualities and colours to make his painting, Bayes, to present and solve his own problem, rejects the use of a lottery framework,[19] which would imply that events can be summed and stored in the same way as countable things. To frame his problem and its solution, Bayes picks instead a square table which is the target for the throwing of balls, the impacts of which are recorded. This throwing that *will be performed* could hardly be understood as *a drawing of balls or tickets*. In the latter case, the content of an urn can stand for the potentiality of all the drawings. This can not be the case for the table that receives the impacts which represent events. Just as there are names which, when given to concepts to designate them, draw the readers further away from and impede their understanding at some moments, and at others draw them closer to it and facilitate it, so there are tales which lead one astray, whereas others, at least for some time, are close allies of the conceptual operation, and collaborate with it and enhance its success in an irreplaceable way.[20] Mathematical problems are necessarily underlain by frameworks which are not properly speaking mathematical, but which can mislead or guide whoever tries to solve them. The difficulty is that if one wants to solve problems one cannot avoid the use of such frameworks, any more than one can avoid using words to refer to concepts. It is because of its anchoring in another and a different symbolic universe and in another imaginary universe that mathematics seems, by means of its own symbolic imaginary world, to know so much about reality.

Failing to select the right mathematical framework – that is, the appropriate rational methods – to accompany its concepts would not be much of a problem if the error had no practical consequences. However, the recourse to calculations about the future by way of the analogy of a lottery induces important practical mistakes, and even serious ethical ones. Not even the best and best-established knowledge in

[18]What is at stake is to know – when one does not have much information about events or subsequences of events – what the chance of being right is when one grants them some probability that is arbitrarily situated between two degrees. Bayes expresses it the following way: ‘Given the number of times in which an unknown event has happened and failed: Required the chance that the probability of its happening in a single trial lies somewhere between any two degrees of probability that can be named’ ([5], p. 26).

[19]As imprudently used by Richard Price, who presented the work of Bayes [5] – who died before finishing his *Essay* – before the Royal Society. The lottery framework cannot mislead once one has the other framework, that of the table and the ball. However, being too little concerned with its singularity, it could not have allowed the invention of the solution to Bayes' problem.

[20]Galileo had that beautiful word *storia*, story, to refer to that type of narrative which stages a conceptualisation. Quoting from *Dialogue Concerning the Two Chief World Systems* [13], F. Balibar reminds us in *Galilée, Newton lus par Einstein*, ([4] note 21), that the *storia* at the time of the Italian Renaissance referred to a painting or a sculpture in several episodes; for example, the doors of the baptistery in Florence.

statistics and probabilities we possess today makes any difference to the strength of this point: the doctor is guilty of a fault towards his patient whenever, by misinterpreting the one or the other, he gives them 'false hope'. Hope consists in presenting as a happy event that has some chance of happening – recovery, for example – something which can only be calculated with a small chance of being right, and consequently with a large chance of not being right, when considering a case which presents particular circumstances. Prudence is probably better than hope, about which Spinoza quite legitimately used harsh words in *Ethics* [23],[21] and which too many carers insist upon, as if it retaining were necessarily virtuous, even against all reason. The doctor does indeed have the right, like everybody else, to 'have reasons to believe', but it is only with the utmost prudence that these should be mentioned to his patient. Going immediately from the collective aggregate to the singular, or from some particular singular case to the aggregate of cases, is always a mistake, but there are, in medicine, which is not only a theoretical work, some mistakes that a doctor may not make without at the same time being guilty of an ethical fault through recklessness.

5.6 Conclusions and Perspectives

1. We have examined a few figures of the mythical – obviously quite briefly and by no means exhaustively. It is well known that some of them hinder research while others facilitate it, at least over some segments of conceptual courses, while they may become obstacles on other segments. That means that they are, at the same time, both invaluable catalysts and unavoidable obstacles,[22] and it is by no means easy to provide the rules governing the transformation from the one into the other, nor even to know if there are any. What is clear is that the investigation can fruitfully be pursued in ethics by looking at Greek myths. G-G. Granger tried to classify styles in the demonstrative endeavour of mathematics [14]. What he called the style of a demonstration – Euclidean, Archimedean, Cartesian, Arguesian, Bernoullian or Bayesian – is not unrelated to the mythical symbolism which underlies it. Similarly, the valorizations which are the basis of ethics are not without relation to the mythical either: why should the continuous be more valuable than the discontinuous? What sympathy attaches to the continuous and serves to discredit the discontinuous in the way we have noted? How are the Promethean values of medicine articulated in relation to the Epimethean values

---

[21]'Emotions of hope and fear <*Spes et Metus affectus* > cannot exist without pain. For fear is pain, and hope cannot exist without fear; therefore these emotions cannot be good in themselves, but only in so far they can restrain excessive pleasure' (*Ethics*, Part IV, Prop. XLVII).

[22]In Appendix VIII of *Chrestomathia*, Bentham [6] showed that the nature of the accompaniment that the myth is in relation to concept is only segmentary. The theory of fictions, born on the neck of Pandora's phial, has always remained a theory of edges or limits: 'how is fiction produced?' and 'how is it removed?' are its main questions.

of care? Why should augurs never be believed, and why do the gods only amplify the successes and failures that men create by their own efforts or clumsinesses?[23] Once again, the rational thought is not always opposed to the mythical one, for it could not exist without its support: the latter readily serves as a framework for the former in an inextricable intertwining, which is all the more dangerous in that its transformation into an obstacle is insidious.

We have seen that the 'core' function of the framework does not have the same role in mathematics as in myths, and does not have the same relation to the symbolic folding, but we could add another argument in support of the comparison between, and even of the joining of, the conceptual – mathematics being a sort of privileged expression of it – and the mythical, insofar as, as Aristotle underlined in his *Poetics*, myths are neither less general nor less necessary than other narratives which apparently do better in meeting that double requirement. In any case, poetry and tragedy, of which myth is the soul,[24] are worth more than the specificity of facts with which the historian deals. Indeed, 'the real difference between a historian and a poet – that is, a man of myth – is that one tells what happened and the other what might happen. For this reason poetry is something more scientific and serious than history, because poetry tends to give general truths while history gives particular facts. By a «general truth» I mean the sort of thing that a certain type of man will do or say either probably or necessarily <κατὰ τὸ εἰκός ἢ τὸ ἀναγκαῖον>; that is what poetry aims at in giving names to the characters. A «particular fact» is what Alcibiades did or what was done to him.'[25] One will have noted that the meeting point between myths and the disciplines which use general concepts lies in the opening of the probable and the future.

2. Greek myths are still of interest for us, and we are not capable of despising them or setting them aside, for at least two reasons. The *first* is, undoubtedly, because they have existed over a very long temporality, which can cross millennia, but to which it would be a mistake to give some archaic value, although they seem to accompany our ways of giving form to reality almost naturally. Foucault warned us against this error, asking us to keep a cool head in the face of that supposed archaic nature, by keeping in mind its historical nature, even though it is a 'long' history, as historians have habituated us to say. One can see here how dangerous the Aristotelian distortion of history and myth is. The *second* is, no less undoubtedly, because we have our own myths which make us feel, by comparison with those of the Greeks, that they are in fact more dangerous than theirs, and even that they could lead to worse abuse precisely because we do not always acknowledge them as myths, and thus grant them a truth value they do not have. We often still allow ourselves to be deceived by the idea of *progress* which is still presented –

[23] In that sense, one could relate the movements of gods to those of stock exchanges which, on the market, in the modern and contemporary era, worsen disappointments and exaggerate successes.

[24] As Aristotle recalls in *Poetics*, ([2] VI, 1450 a 20).

[25] *Poetics*, ([2], IX, 1451 b 1–10, p. 35).

though admittedly less and less often – as a real certainty: have I not said at the outset that our medicine is more efficient than that of the Greeks, that we know the mechanisms of illnesses better than they did, and, more generally, that we know more than they did? This idea of *progress* has also imposed itself by a strange *coup de force*, without any demonstration or rational argument: how can we reduce to the same common denominator vastly different activities in order to permit ourselves to say that an activity is truer, fairer, more useful or better than another? A more or less naive spirit of unification, rectification and measurement inheres in and permeates the idea of progress from which it is difficult for us to break away, precisely because it legitimizes the common measures we need, or think we need.[26]

3. Thirdly, we must pay the keenest attention to the affective frameworks, which are full of images –and which are at best irrational – which accompany rationalities. We would like, in this respect, to return briefly to a bias that sometimes exists among the historians of probabilities, and which seems flagrantly wrong. We were taught, some 20 or 30 years ago, that the calculation of probabilities was made possible only because of some economic conditions linked to liberalism – a certain way of exchanging, inheriting, enterprizing, of considering capital and of sharing it – and that, consequently, the Greeks were not even capable of thinking that it was possible to calculate on the basis of what was probable. Admittedly, this much may be safely conceded: although they showed themselves to be astute logicians in relation to the probable, the Greeks did not calculate probabilities, though they corrected curves, and determined quadratures and centres of gravity. However, the assertion that liberalism shaped the calculation of the probable is only understandable from the moment when the mathematicians started to calculate the probable by developing the habit of using frameworks of games or money, or even of both at the same time. That monetary framework is not simple rhetoric. It is quite possible that such 'monetization' very tangibly inflected the issues of probability, even though the mathematicians, coming from societies with very different economic systems, all approved solutions which yielded to the monetary *coup de force* which both insists on and depends on the commensurable nature of qualitatively different realities. However much in love with money the Greeks were,[27] they did not reduce it to different probabilistic estimates, and they never thought of bringing them closer in any other way. However, the lesson we can draw from *The Persians* or from fragments of Diodorus Cronus [11] is not that they lacked liberalism, but, conversely, that we might draw our inspiration from a 'demonetized' way of thinking about probabilities, and look at what would happen to the idea of the probable – what extension and import it would have – if one pursued its 'demonetization' as far as possible.

[26]What we are saying here about the idea of progress could very well be said of many other ideas to which we are equally attached in quite a mythical way: Democracy, human rights, the federalism of European States, and many other values fall into this category.

[27]Mauss stresses this feature in *The Gift*. [17]

4. The reader may have been struck by the fact that we have not explicitly used, in our remarks, the mythical names that psychoanalysis has provided since its foundation by Freud, for nodes of affects that the psychoanalyst rationally analyses in addition. Everyone knows of the Oedipus complex and the Electra complex, of Antigone, who is discussed at length by Lacan in *The Ethics of Psychoanalysis* [16], of Pandora's box – to which we have but very briefly alluded – and of the everlasting conflict between Eros and Thanatos. We have not used them because psychoanalysis itself, in some of its versions, is reluctant to be understood as care, and we are fundamentally dealing with medical ethics. However, its influence on this work should be recognized if only because, in quite a marked way, it makes us think that the names we give to things have a logic that is different from that of the concepts or ideas to which they are attached, and thanks to which we conceive of those very ideas in the way in which we do. Moreover, the myths through which psychoanalysis conceives and articulates itself tell a story that is in counterpoint with rather than identical to the associated concepts or demonstrations that are nevertheless intimately attached to that story. The cleft or bifid nature of stories – or even trifid, quadrifid or n-fid division of stories – is unavoidable, not only in human sciences, but everywhere that concepts must be named, which is to say everywhere. One cannot think without words, but that thinking in words brings with it serious ambivalences that the non-specialist who learns to conceptualize them must always confront as an obstacle, whether they are aware of it or not. The reader will not be surprised to hear that, in my own view, it is better that they should be aware of it. The very names and sentences we use to refer to concepts are the myths of those concepts, until those concepts themselves become, in their turn, mythical when expressed in another science.

If it were objected that this point is quite trivial or secondary, and holds almost no interest except for the philosopher or the specialist, I think the objector would be making a serious mistake. To support this assertion and to finish, I would like to give the reader one more reason in support of the argument of this chapter. The same concepts do not have the same names in all languages, because not all languages mythify, or perhaps mystify, in the same way. However, in the same way that French-speaking people desired, during the Empire, to impose their measurement system on the whole of Europe, and almost everywhere in the world, English-speaking people – who speak a language which I think is the language of a great culture – have for several decades now been busily seeking to limit and reduce the possible ways of denominating illnesses and their remedies in other languages, on the ground that they have no value but in relation to their own language, with its own imaginings, its own myths, its own symbols, and its own access to reality, in a process in which the 'good reasons' of Evidence-Based Medicine play a central justificatory role. However, this operation does not involve only 'good reasons': there are 'bad' ones too. To mention only one problem, it can be very difficult to denominate one's illness in a language that is not one's own. Here again, we can hardly deny that what has been at stake in our problem is not only theoretical, but has important practical, and therefore ethical, value.

5. Lastly – and this concluding remark will be our link to the following and final chapter – the link between the myth and its commentaries, like that between a theorem and its supersession by a more powerful mathematics, belongs to a sort of fold in the symbolic for which it is possible to account with the terms of a theory of fictions. This may be – and we have already suggested it in this book – one of the rare reproaches that could be levelled at Hare's ethical philosophy, that of having admitted and distinguished between only two levels of ethics, and of understanding their possible inversions, but of not multiplying either the intuitive - or so-called intuitive– or the critical level into a series of levels which would have rendered the process historical. Hare mostly ignored the theory of fictions, which he seems to use in some moments, but never in a systematic way. The proper moment of fiction or fictionalization is when an entity or a level of entities that are thought to be real produce a whole range of imaginary or symbolic entities which seem to assert their reality in other modes, while accepting a process of reversibility, so that what is held to be real may, from another perspective, be thought to be fictitious. We have shown in other books how perception, like the passions, was likely to be characterized by the rationality of the fold, which is so important in all the issues in which human affairs, and consequently, ethics are in question. We shall demonstrate it once more in dealing with the issue to which we turn in the next Chapter.

Bibliography

1. Aeschylus (1988) *The Persians*, in *Aeschylus*, trans. H.W. Smyth, 2 vols. Harvard University Press/W. Heinemann, Cambridge, MA/London, vol. I.
2. Aristotle. (1991). *The poetics*. Cambridge, MA/London: Harvard University Press.
3. Aristotle. (1989). *Aristotle in twenty-three volumes*, XVII, *The Metaphysics* (trans: H. Tredennick). Cambridge, MA/London: Harvard University Press/Cambridge University Press.
4. Balibar, F. (1986). *Galilée, Newton lus par Einstein*. Paris: PUF.
5. Bayes Th. (1988). *Essai en Vue de Résoudre un Problème de la Doctrine des Chances*, Cahiers d'Histoire et de Philosophie des Sciences, n° 18, ed. Belin, Paris.
6. Bentham, J. (1993). *Chrestomathia* (eds: M.J. Smith & W.H. Burston). Oxford: Clarendon Press.
7. Bernoulli, J. (1975). *Artis conjectandi, pars quarta, tradens usum & applicationem praecedentis doctrinae in civilibus, moralibus & oeconomicis*, in *Die Werke von Jakob Bernoulli*, Birkhäuser, Basel (3rd vol., pp. 239–259).
8. The Bible, *The Gospel according Saint John*, in *The New Testament*, King James version, The Gideons International.
9. Coumet, E. (1970). La théorie du hasard est-elle née par hasard?. *Annales Économies, Sociétés, Civilisation*, 25th year, May–June 1970, pp. 574–598.
10. Desargues, G. (1951). *L'oeuvre mathématique de Desargues*. Paris: PUF.
11. Diodorus Cronus, in: Vuillemin J., 1984, *Nécessité ou contingence, L'aporie de Diodore et les systèmes philosophiques*. Paris: Les Éditions de Minuit.

12. Emanuel, E. J., & Patterson, W. B. (1998). *Journal of Clinical Oncology*, official journal of the American Society of Clinical Oncology (vol. 16, N° 1, January 1998, pp. 365–371).
13. Galileo Galilei (1967). *Dialogue concerning the two chief world systems* (trans. S. Drake). Berkeley/Los Angeles/London: University of California Press.
14. Granger, G.-G. (1988). *La philosophie du style*. Paris: O. Jacob.
15. Hilbert, D. (1971). *Les fondements de la géométrie*, Paris, Dunod. 1930, *Grundlagen der Geometrie*, ed. Teubner, Leipzig.
16. Lacan, J. (1986). *L'éthique de la psychanalyse The ethics of psychoanalysis*, in Le Séminaire de Jacques Lacan, éd. du Seuil, Paris.
17. Mauss, M. (2016). *The gift*, selected, annoted, and translated by J. Guyer. Chicago: Hau Books.
18. Passamani, E. *The New England Journal of Medicine* (vol. 324, N° 22, 30 May 1991, pp. 1585–1589).
19. Plato. (2013). *Republic*, Books 1-5, ed. & trans. C. Emlyn-Jones and W. Preddy. Cambridge, MA/London: Harvard University Press.
20. Plato (1984). *The Laws I & II*, in Plato in twelve volumes, X & XI. Cambridge, MA/London: Harvard University Press.
21. Sextus Empiricus. (1987). In four volumes, III, *Against the ethicists*, trans. rev. R.G. Bury (pp. 383–509). Cambridge, MA/London: Harvard University Press/W. Heinemann/The Loeb Classical Library.
22. Sicard, D. to President Hollande on 18th December 2012, and was entitled *Penser Solidairement la Fin de Vie Thinking about the end of life in solidarity*.
23. Spinoza, B. (2013). (?), *Ethics, (ethica ordine geometrico demonstrata)*, trans. from the Latin by R.H.M. Elwes, globalgreyebooks.com.
24. Villani, C. (2012). *Théorème Vivant*. Paris: Grasset.

Chapter 6
Medicine, Apparatuses, Robots and Intimacy: A Few Ethical and Political Aspects of the Linkage with Machines

Abstract Fortunately, far from J. Ellul and from M. Heidegger, contemporary ethics invoke no more humanism or religious convictions in order to reject technologies, as if they were hostile to man. We have learned and are always learning to live with machines on which, very often, our health and even our very life depends. The existence of men in flesh and blood and the being of machines are intermingled in such a way that it is no longer possible to discern where the one starts and where the other ends. Yet, if this linkage between men and machines is generally recognized, without being rejected, the ethical problems continue to exist as if men had theirs and machines -digital machines and « intelligent » machines- had their own ethics or raised unique ethical concerns. What is quite absurd and opens wide the door to trans-humanist and post-humanist myths that cannot escape the qualification of *Schwärmerei* following the appellation that Aufklärung gave to such elucubrations. Nevertheless, this is not the issue: ethical questions deal with the indefectible mixed being that any modern man, member of our cultures, makes up with machines. A certain number of political conceptions are disqualified when these linkages with machines are taken into account; we highlight the point with Rousseau's contractualism. But it may happen that moral positions, even classical, like Kant's moral law, do not need to be applied to human beings. At least, Kant has fancied the first of the three formulations of his moral law for beings that are not specifically human. The matter is not to substitute Kantian morals to present ethics but to highlight that the ethics we are searching has predecessors and that ethics, when human nature is not imagined to be steeped in laws, cannot avoid dealing with linked or coupled beings.

Keywords Artificial intelligence · Benthamian conception of law · Contract · Democracy · Ethics and politics · Intimacy · Linkage of human beings and machines · Merleau-Ponty M. · Robots · Transhumanism

Insofar as there are very few disabilities that do not result in a proximity and an interdependence of the disabled with machines – except perhaps, and maybe not for that much longer, for mental disabilities – I think it is a good idea, in an effort to

J.-P. Cléro, *Reflections on Medical Ethics*, Philosophy and Medicine 138,
https://doi.org/10.1007/978-3-030-65233-3_6

think about disability, to begin with a few thoughts about the link that medicine has established between an increasing number of patients and all sorts of increasingly sophisticated apparatuses, machines and robots. We would like to assess, from the ethical point of view, this linkage that can be extremely diverse between the patient and the different types of machines to which he entrusts an important part of his life, and sometimes his whole life, in the detail of its duration, and without any possible relief. Is a life which is intimately dependent on machines, whatever their appearance and way of operating, worth living? Is it an alienated life, a life that has become a stranger to itself, in that strange alliance, as if there were a self to which it were possible to refer beyond that alliance? A life which, though it were improved in some important ways, would pay heavily for the weight of its linkage with machines? We would like to analyze the idea of *intimacy* that we developed at the end of our last book, and revisit it at precisely the point where it concerns the couple of a human being and a machine, and when the latter plays a fundamental role in the mode of existence of the former, and even in the continuing possibility of its existence.

Even if everyone will agree that intimacy is not devoid of value in medical ethics – provided it is adequately defined and not left, without being specified, in close contact with other categories which tend rather too much to annex it completely – it will be faced with a major obstacle which seems to threaten it, and even to condemn it. The robotization of medicine, which follows at least two paths – that, *on the one hand*, of the refinement and increasing safety of operations, and, *on the other hand*, that of the linkage or hybridization with machines which not only help the patient who could not live without them, but which are also likely to improve the performance of the human being, whether they are ill or not, in quite diverse areas. The first case belongs to a classical ethical setting, despite the fact that trust towards human beings is combined and mixed with trust in the reliability of the machines on which the patient depends, while the second case lies much more obviously outside the classical framework, and has resulted, because of the quite radical decentering to which it submits the subject or the individual – because of its hybridization with machines – in transhumanist or posthumanist speculations.

Thus, in the face of the idea of *intimacy*, robots may seem, from the point of view of those who want to link the former to the latter, incongruous partners. Is the capacity for intimacy not precisely what these machines lack, even though they were granted artificial intelligence, a tremendous 'memory', and extraordinary capacities for the greatest precision in tasks, for predicting the likely consequences of actions, and for not only storing data, but for using it in constructive ways? It is well known, from having been repeated so often, that machines do not have affects, personality, will or self-consciousness. However, even if we admit such commonplaces to be, at least in part, well founded, we would still like to show that *intimacy* retains some meaning and plausibility in the linkage that 'human' individuals may form with robots, and that that confrontation even makes it possible to draw distinctions that are not usually so refined. One may oppose, as Cartesians do, the machine and human intelligence. Human intelligence, when coupled with the machine, is not the same intelligence (or the same individual) as before. It is also possible to oppose, as Kantians do, natural and artificial things and persons ([21], p. 79, p. 378). Then though, one must realize that *intimacy* is not the same idea as *person*, that a person does not necessarily identify with the intimacy of the being they represent, and that the *autonomy* which characterizes the person does not necessarily require any *intimacy*. It is quite possible, on the contrary, that a machine that is coupled with, or connected to, a man, constitutes quite a specific form of intimacy.

We are going to give a quick outline of the reasons why *intimacy* cannot be reduced to any of the traditional categories of 'personalistic' ethics. Before that, however, we need to note that, even though the increased robotization threatens to undermine or even erase the meaning of the ideas surrounding the *person*, *personality*, *consent*, and *enlightened consent*, these ideas have in fact been losing their meaning for a long time, as much because of external developments in the environment in which they are applied – besides being irrelevant when they applied in unpropitious conditions, for example when we talk about the consent to treatment of certain mental health patients who are not asked for their opinion – as because of their internal contradictions. In particular, the threat to their meaning has come from two main sources. These are on the one hand utilitarianism, which implies a calculation of pleasures and pains, or a calculation of preferences, and on the other the deep re-examination of the idea of *trace* as Jankélévitch developed it, ([20], chap. V) which moves beyond the *person* and the cluster of ideas linked to it. Those two perspectives, of utilitarianism and of Jankélévitch, which seem so far apart from one another, are not in fact incompatible, although, as we showed by other methods in our first book, ([10], pp. 139–155) their reconciliation is problematic and their articulation in relation to each other requires much more work.

The paradoxical result of this analysis is that *intimacy*, which seems to be one of the most delicate and fragile ideas in ethics, may actually be in better health, and may prove to be more viable than a great number of other notions the decline of which is accelerated by robotization – quite simply because their erosion no longer proceeds by means of theoretical philosophical and historical discourse or dialectical discussion, which undermines and weakens ethical categories from the inside, but through the practical test, which is unforgiving and irrevocable, of reality, which confronts us, with an urgent demand for serviceable answers before the *fait accompli* of experience which, far from permitting us the luxury of detached and reflective observation, draws us willy-nilly into the adventure.[1]

Who would oppose brain implants, even though they seemed to threaten our individual integrity or personality, if they allowed us to gain the advantages of a sensitivity that we have never enjoyed, or to recover important capacities which we have lost? How could we not view positively the work of the Caltech team in the United States, which has allowed a tetraplegic to drink on his own, by raising a glass of water to his mouth thanks to a system that uses the areas of the brain that are mobilized by the intention to act, thus artificially translating the wish alone into actions which constitute the execution of the task desired? If these examples are still

[1]Everyone will have recognized, in this perspective, the idea G. Hottois defended in *L'Inflation du Langage dans la Philosophie Contemporaine* [17].

anchored in the therapeutic field, one can speculate that it may be possible to improve and multiply our capacities by progressively abandoning any reference to the therapeutic, and that there may be few arguments available for rejecting that path, even though losses were associated with the gains. Rather than speaking in terms of the discourse of the undesirable intrusion of the machine into our personalities, of the violation of individual integrity by robots, and of the *alienation*[2] that the dependence on these products of artifice represents for our being, we must look for a connection of *intimacy* with the machines to and with which we are to be coupled or hybridized, and distinguish it from the relations of *person*, *autonomy*, the *private* as opposed to the *public*, *interiority*, *individuality*, and even *humanity*, for we will argue that intimacy transcends the very concept of *humanity*, and that, even if the notion of *humanity* lost its meaning and its value, the notion of *intimacy* would direct us to retain it, and would resist its disappearance as much as it would that of the concepts that we were mentioning earlier if they were to be jeopardised. It behoves us to try to understand why this should be so.

6.1 The Confrontation of *Intimacy* with Neighbouring Concepts When One Seeks to Attribute Them to the Machine

Nobody will say that a machine – even an intelligent one – has a personality. When it is coupled with a human being, it is the latter who is given the benefit of that concept, out of a sort of instinctive dualism. For a machine to have a personality, it would have to be able to project and identify itself as a machine, and even as a machine with as yet unrealized potentialities which it might and should develop.

If the concepts of *person* and *personality* are already phantasms when used about a human being who is sane, and who can project the difference between what he is and what he might be, they become quite absurd when applied to a machine which enjoys autonomy only within certain, more or less tight, more or less random, limits. It is quite certain that it is possible to design adjustments of machines according to programmes that include a margin of probability, but machines do not invent their own programme. For the moment, it is the IT engineer who, by designing or tuning it, provides for something of that order, even though it is possible to imagine that he is dependent on his own creation – even though that phantasm of dependence were necessary to a new creative momentum for that matter. It is not absurd, but quite the contrary, to speak, at least in some circumstances, of *intimacy* with the machine, or of the couple of the *human being* – if one is still content to use that term – and the *machine*. We are all familiar with the experience of a machine which used to belong

[2]We understand that word no longer as an *exteriorization*, which the Germans readily call *Entäusserung*, but as a process which renders us foreign to ourselves, and which is expressed in German as *Entfremdung*.

to someone, and which remains 'hers' long after death has partly severed that link, perhaps through a mechanism of projection, but not only as some 'hau'[3] – the famous 'hau' of the New Zealand Maori society that Mauss studied – which belongs to the object itself and resists the breaking of that link.

We will note here that a thing, which does not have a personality and is usually opposed to those beings which do have one, retains an intimacy, even if a man did not contribute much, or at all, to its existence. It makes very little difference, indeed, for intimacy as we conceive it,[4] whether the individuals who give the machine its form are human beings, or whether that machine belongs to their system or not. What matters is that intimacy is engraved into existence, more deeply than any concept of conscience and will, that it cannot be taken away from any individual or from any linkage between individuals, and that the individual himself, even when that engraving is his, cannot exclusively determine it since it is in part bestowed, and will always be bestowed, on him, even when he approves of it and desires that it should be bestowed. For even though the partial mastery we have of that engraving is part of its phenomenon, it is not its essence, and nobody masters everything that can be engraved into being as signifying or bearing meaning. We have at best some mastery over the meaning of the inscription, but such mastery is quite fragile in that what an individual manages to engrave of his existence does not coincide exactly with what he wants, goes beyond what he wants, and is not correctly identified with what he wants.

6.2 Difference Between the Ethics of Individuals (or of Intimacy) and the Ethics of Persons

What difference does it make in ethics to rely on individuals rather than on persons, and to try to give the most important space to the concept of *intimacy*? To answer that question, one must confront Kant's morals.

In the second of the three formulations that Kant provided of the categorical imperative, he introduced the concept of *humanity* and that of *person* in the same sentence ([21], pp. 79–81), which makes it more ambiguous than the first formulation, which only requires – without going into either the detail of what should be called *humanity*, or the inherent difficulties and contradictions of the concept of the *person* – the elevation of the maxim of the agent's action to a universal law (of nature). Admittedly, the concept of *nature* is also ambiguous. We will not tackle the issue of which ambiguity is more problematic, the first, which belongs to that nature which coincides with our construction and articulation of laws, or the second,

[3][26], pp. 26–30. The 'hau' is that strength which is in the given thing and which forces the beneficiary to give it back under one form or another. Things, themselves endowed with a spirit, oblige the beneficiary.

[4]For a more complete analysis of intimacy, [10] pp. 139–160.

which points to us as thinking and willing beings, whose ends are superior to those of other beings.[5] The author of *Groundwork of the Metaphysic of Morals* does not make things easy when, in the second formulation of the categorical imperative he adds a 'human' component to what should be taken into account in ethics. Kant did not necessarily think of reason and the law it gives to itself in ethics as essentially human,[6] and in this at least he endorses our own approach. We will ask in the same way if one is really speaking more clearly when one talks of the 'human body'. How is the body 'human'? If, for example, it is coupled with a machine, does it become 'inhuman'? Is it important in any way to call it 'human', and how should that designation be upheld and maintained as a marker proper to ethics, above all if one adds, as Kant does – without apparently fearing circularity – that if humanity is sacred in the person of man, that is because he is the subject of a moral law, which is sacred, thanks to the autonomy of its freedom?[7]

The problem in the linkage of the body with the machine(s) which makes of them both taken together a new composite body, is to recognize and to discover what it is that is thus engraved. Does the linkage of the body with the machine, imagined in terms of an essence and with limits that the new compound does not possess and which do not have any reality, pose problems which are so very different and really more serious than those posed by the body which I imagine as being 'mine', or which I take as being 'mine'? I am not my body. I am no more my body than I am the phantasm of the linkage I form with the machine that helps me to live, and which transforms me by helping me to live. That phantasm is – like the previous one in which I believed that I was my body – the expression, the transient, unstable expression, of something that demands more, that exists more, which is therefore much more 'real' – which does not prevent it from being elusive – or that something which is only understandable in terms of those very phantasms which unavoidably betray it. What is being written? Why will one such compound composition be written rather than some other? This is the question of *intimacy*, which is easier to deal with through apagoges, or indirectly through negations, than directly through positive assertions.

It is obvious that what is being written is not simply a pre-existing 'me' equipped with a machine auxiliary. The combination, the compound, is often more radical than the simple addition of an appendage. As to the 'me' that would like to understand itself apart from and beyond the mechanical, through the instinctive dualism that I have mentioned, it is only imaginary and phantasmatic. What is being

[5][22], p. 401. The intelligible character of the transcendental subject 'nous est indiqué par le caractère empirique comme par son signe sensible'.

[6]A. Philonenko insists on this in his analysis of Kant's practical philosophy: 'Sans cesse, dans ses textes moraux essentiels, Kant parle de l'être raisonnable et non de l'homme' ([37], p. 25). Kant always speaks of « reasonable being » rather than « man » or « human » in his practical philosophy.

[7]*Critique of Practical Reason*, B. I, Part. I, Chap. III ([21], pp. 198–211). One really does have the impression of going around in circles. It is this circle that a certain number of thinkers like Foucault, Levi-Strauss and Lacan wanted to break in the middle of the 20th century (See Lacan J., [23], pp. 369–370).

written of the coupled-being is an enigma, which is constantly problematic. It is at the same time the most fragile and the most resilient of the phenomena of existence. Let us not specify that it is the existence of the self or of the person, for the sturdiest and the most particular aspects of existence are not contained or signified in those phantasms. The best or more fruitful ways of reading or interpreting those aspects are neither immediately self-evident nor easily demonstrable. At the very least their immediate reading is as precarious as that which may be made later. The precariousness of such a reading, however, does not erase the very hard chiselling of the writing of it.

To understand the writing or the chiselling of a thinking being coupled with a machine, we need to stop understanding that being as a thinking being, or even as a being possessed of a power of synthesizing its representations, which is at the same time opposed, associated or united with the extended thing that the machine is. Machines have not been characterized by extension for a long time, and the Pascalian way of conceiving the self, which has been so extensively criticized by Cartesians and Kantians for its supposed inadequacy, seems to us on the contrary remarkably open to the possibility of linkage with all sorts of objects – among which are the 'intelligent' machines[8] that Pascal started to invent and think about. Everybody knows the fragment of *Thoughts* which is entitled 'What is the self?', in which Pascal, remembering Montaigne's *Essays* [30], twice states that 'we never love anyone, but only qualities', and 'we never love anyone except for borrowed qualities'. Even if one thinks one loves someone for what he is and thinks one is loved for what one is, the self is only a set of motley qualities. It is then that it becomes quite admissible to believe that among those qualities are included the types of qualities which characterize the objects that machines can be, as Montaigne ironically suggested.[9] It is Montaigne's irony that needs to be challenged and overcome here.[10]

The writing of existence makes use of heterogeneous phenomena which have physical, chemical, organic, mental and psychosocial characteristics or qualities. It is this extraordinary impurity of the phenomena through which it is written that allows room for the linkage with the machine. It is also why it would be absurd to exclude the machine from that compound-being, or to make it merely secondary as a sort of appendage, as if it were true that the being that exists is the result of the linkage of a human being with a machine, and it were possible to identify the human being. Such writing is deeper than the concept of *person* and its associated concepts, which we usually take too seriously given their obvious infelicities, which mean that they have never worked unproblematically, and now, as we have seen, will work even less

[8]I am referring here to machines that perform acts which are normally expected of intelligence: calculation was for a long time among them. Had Plato not seen in it what constitutes our humanity, more essentially and more archaically so even than language, in *Epinomis*, 977c ([38], p. 439)?

[9]In Book I of *Essays*, Chap. XLII, ([30], pp. 254–263).

[10]'Qu'on ne se moque donc plus de ceux qui se font honorer pour des charges et des offices' (Sellier, 567, in Pascal B., [36], p. 1135). «Let us stop scoffing at those who win honour through appointments and offices» ([35], p. 218).

well. If what is being written about existence is probably linked to consciousness and will, it is yet much deeper than both, and it remains the case that it is from that inscription, deeper than the ordinary bodies of ethics, that one should try to construct a valuable ethics, without starting from the concept of the *person*, which of and in itself would make an ethics for present times – or which can accommodate what may be tomorrow – quite impossible.

It seems to us, as we have already suggested, that the best existing draft of the logic of the individual that we are looking for is found in the second book of Spinoza's *Ethics*, [47][11] which comes before the book on passions or affections and which provides the key to their composition. First, there is that chiselling, that inscription in being. Then there is a logic of the composition of bodies, which is also one of minds. The composition of bodies one with another are the signifiers which most of the time we live and think of only in phantasm and imaginary modes. Quite curiously, there is enough here to reconcile Spinoza with Jankélévitch – two thinkers whom it seems risky to bring together. Jankélévitch is the thinker who, in modern times, was maybe the best able to give a voice to the inscription we are seeking.[12] We know the historical events that led him to do it,[13] but we should emphasize the dimension in his philosophy that transcends particular circumstances, though those in which he wrote were quite extraordinary because of the paroxysm to which Jankélévitch's ethics was intended to supply a response.

6.3 The Engraving of the Ethics of Intimacy

Let us explain what we understand by this engraving. It is produced by – and at the same time not necessarily produced by – real writings, images or photographs, even though it presupposes them. It is rather like the accompanying curve that is present in

[11]And, maybe also in the logic of *liaisons* and *déliaisons* which are to be explicitly found in Pascal and – more implicitly – in his sporadic reflections on machines.

[12]*L'irréversible et la Nostalgie*, [20], p. 337: 'L'avoir-été, qui nous apparaissait en creux ou à l'envers comme une inconsistance, va nous apparaître en relief ou à l'endroit comme un acquis inaliénable et un gage de permanence: car si l'irrévocable de l'avoir-fait est relativement irréversible puisqu'on ne peut le revivre ni même le refaire, l'irréversible de l'avoir été est à son tour relativement irrévocable puisqu'il ne peut plus être nihilisé. La positivité négative de l'avoir été, tout comme la dissymétrie de notre demi-pouvoir, atteste l'intermédiarité de la condition humaine, et elle est une assurance contre le néant. Nous parlions du caractère, sinon indestructible, du moins inexterminable du fait-d'avoir-eu lieu en général'. And, later: 'Entre le *non-être* et *n'être-plus* il y a toute la distance infinie de l'avoir été; et rien au monde ne peut plus désormais faire que celui qui a été puisse ne pas avoir été: désormais ce fait mystérieux et profondément obscur d'avoir vécu est son viatique pour l'éternité' ([20], p. 339).

[13]One knows that the Nazis wanted not only to commit their crime against the Jews, but that they even wanted to delete their crime itself. Hence the necessity to elaborate an ethics of the engraving into being.

the problems of classical mathematicians, or like the line of music that runs through musical works, those of Schuman, for example,[14] which can be heard, or rather listened to, without allowing itself be empirically noted on a score or embodied in it, or like the line A. Bosse mentions, which runs on and ramifies in monuments, and to which the stonemason implicitly aligns himself when he reads and puts together the architect's plans. The paradox is that the signifiers, however robust they may appear, are not present properly speaking in any experience, whereas the signified, however elusive and friable they may be, largely occupy the field of experience.

Then there is that logic according to which one individual, by being linked in divers modes to others – a machine may be for this purpose be considered as an individual – becomes another individual, without us being able to suppose that the changes in an individual are changes in an identical subject through the modifications that affect him because of his link to another being – be it a body or a mind. He remains the same individual only upon some conditions, and may become another individual by being allied, in different modes, to other beings or individuals. According to this logic, a break or a change in the necessary conditions for the maintenance of the body – let us call it 'human', if we like – is not problematic.[15]

Man has never existed as any sort of eternal nature. He is only what he has been fixed or defined as – what he was fixed or defined as, even – under that term for some particular period, some centuries, to borrow the measure that Foucault gives of it.[16]

[14]We are referring here to Schumann who, in *Humoreske of Klavierwerke* IV, staged and asked us to listen to an *innere Stimme*, an intermediary voice, between two melodic lines which are really played on the piano, and directly audible.

[15]If the Lemme IV of the Second part states that, 'if, from a body or individual, compounded of several bodies, certain bodies be separated, and if, at the same time, an equal number of other bodies of the same nature take their place, the individual will preserve its nature as before, without any change in its actuality', one at once understands that those replacements could change the individual by increasing or decreasing his powers. This is how, since 'all the modes, in which any given body is affected, follow from the nature of the body affected, and also from the nature of the affecting body', wherefore 'their idea also necessarily involves the nature of both bodies', and 'therefore the idea of every mode, in which the human body is affected by external bodies, involves the nature of the human body and of the external body' (demonstration of Prop. 16). *The Ethics (ethica ordine geometrico demonstrata)*, [47], pp. 51–52.

[16]This Foucaldian "anti-humanism" is particularly sensible at the end of *Les Mots et les Choses*, in the very last paragraph of Chapter X about human sciences: "Une chose, en tout cas, est certaine: c'est que l'homme n'est pas le plus vieux problème ni le plus constant qui se soit posé au savoir humain. En prenant une chronologie relativement courte et un découpage géographique restreint – la culture européenne depuis le XVIe siècle – on peut être sûr que l'homme y est une invention récente. [. . .] L'homme est une invention dont l'archéologie de notre pensée montre aisément la date récente. Et peut-être la fin prochaine" ([13], p. 398). Those who, with the strongest certainty that there exists a man, talked about dealing with human nature or investigating it, knew its frailty very well. Did Hume not say that, if one harmed the principle of causality, one harmed at the same time human nature itself? Behind the appearances of a description, Hume in reality defends human nature [19]. One century before Hume, Descartes hesitated to speak of nature, as he thought that the idea of *law* was enough, and did not need to be supported thus by that of *nature* [12]. As to Boyle, the text that he wrote on nature, *A Free Enquiry into the Vulgarly Received Notion of Nature*, (1686) [8], aimed to show that the idea was dangerous, and that it would be quite possible to do without it

One should not cling to the prevailing figure or definition as if it were timeless, as if it were (and always will be) what it always has been. One can no more know now what one's future is, than one has ever been able, at any time, to know what it would be. If we would understand this logic of engraving, an individual rather than a human inscription, it is from it here that ethics should start, not from founding our enquiry on the nonsense of the *person*, which, however, is not entirely without some foundation of its own, and may still be useful on some points that we will specify later. Thus the true challenge in ethics is not to persist in wondering whether there is something of mankind in any given situation, but rather whether the inscriptions we are talking about – of which we cannot say whether they remain human or not, and even less whether they will in future – are compatible with each other, whether they can combine and intertwine without too many conflicts and contradictions, or in other words, whether a time and space can accommodate them without producing too many reciprocal destructions. How, *a priori*, can the lines that have been written by a robot, or by any individual linked to a robot, necessarily threaten what is written by other individuals, whether or not those individuals are themselves coupled, or not, to robots?

Curiously enough, although one could be ironic about those morals to which Kant seemed to give meaning even though there were no such thing as man, it turns out that it is precisely by not insisting on the idea of *man* in *every* part of his morals that Kant may have succeeded in writing one of those rare practical philosophies, in particular in his moral law, and his first expression of the categorical imperative, which may be applied to the hybridizations we are talking about.

If we prefer speaking of the existence of the linkage of what we are still calling 'man' with the machine, to speaking of a help that would be brought to man by the machines that he creates, it is because the idea of *help* or the function of *auxiliary* that the machines are supposed to bring to and perform for the disabled have become patently insufficient and too inaccurate properly to describe those linkages. The robots and machines of the same type have obviously recently moved to a higher level. Up until now, they replaced us in functions which we performed defectively without their help: they alleviated disability. We were still living under a Pascalian regime, and sometimes even a Cartesian one, with their conception of machines which could perform a very limited number of tasks better than us, [11] although we wanted to multiply what we could do endlessly, even if we could not do it as well as them or as quickly. We expected them to compensate for our shortcomings and satisfy our wants, and, in so doing, to resemble us. We have reached a phase where they do not look like us at all any more and can, by being associated with us, do absolutely unheard-of things.

Just as physics gradually became emancipated – in an accelerated way in the nineteenth century – from the geometrical representation which claimed to give it an

without any damage. The fear Pascal underlined as to the natural feeling of paternity may be extended to whatever can be held to be natural: as soon as man acts towards what is natural, to support it, he admits that he does not really believe in that naturality.

imagined equivalent through its resemblance with the physical and dynamic phenomena it featured, so robotics, albeit much later – the process only began in the last decades of the twentieth century – evolved from an age in which it imitated man, to that of robots which do not have any resemblance to man, and which will probably become emancipated from the very functions they were initially designed to facilitate or enhance. One could – by analogy with what happened in the relationship between physics and mathematics, and in each of these sciences taken separately – talk of a symbolic, and not only of an imaginary, age of robotics, in which it is really becoming independent. The symbolic dimension of analysis or algebra no longer has anything in common with a geometrical representation, but takes experience into account or has some leverage on it in a very precise way, though it no longer imitates it in any way.[17] Nano-robotics, of course, does not try to resemble man in any manner and its performance could not compete with those of 'man'.[18] It is clear that we could avoid a certain number of naive statements on the love we have for robots or on their love for us, or even on their love, or lack of it, for each other, if this new aspect were taken into consideration.[19]

6.4 The Illusions of Merleau-Ponty's 'I Am My Body'

One of the major mistakes in reflecting on of the methods of ethics, and one related to the world into which it plunges us, is traceable to the fact that, like Merleau-Ponty [28], we want to root machines and all symbolic activity in what some phenomenologists have called 'the body' or 'my body', by claiming everywhere – as if it were the key to the foundation of all the symbolic phenomena – that 'I am my body', and by considering this rooting in 'the body' or 'my body' as an original and an ultimate

[17]In his first seminars, Lacan showed a high interest in this type of machines, which he did not oppose to the 'human being', but which, on the contrary, he showed to be commensurable with him.

[18]In *Medical Robotics*, it is written, about robots, that 'pour un bon nombre d'applications de chirurgie, il est pertinent d'intégrer les mobilités et les capteurs à l'intérieur du corps plutôt qu'à l'extérieur. En d'autres termes, plutôt que de manipuler avec un robot extra-corporel un instrument rigide multimillénaire (comme des ciseaux ou des pinces, par exemple, l'idée est de développer une robotique intra-corporelle, offrant au minimum les mêmes performances de qualité de mouvement, de sécurité, d'interaction avec le praticien. Le sujet n'est pas nouveau, puisque les premiers travaux datent des années 1990, mais il est loin d'être épuisé' ([27], p. 394). And, later, ([27], p. 409), 'La robotique chirurgicale évolue vers des solutions déliées, miniaturisées et dotées d'une certaine autonomie. Les capsules ingérables par voie buccale en sont un bon exemple. On peut s'attendre à des développements équivalents avec des dispositifs de taille millimétrique pour amener un médicament ou un capteur sur une cible donnée.'

[19]It was thematized in its principle by Leibniz, when he distinguished clear, distinct, adequate symbolical and intuitive ideas [25]. It is obvious that for him symbolical ideas – which do not imitate their object and do not require its presence – are much more productive than clear, distinct or adequate ones. The more a language severs itself from a strong link with the imagination, the more chance it has of being powerful in its thinking and practice on a reality it does not imitate.

foundation. However, this way of thinking is so far from providing the key to the symbolic that instead we might suspect that it allows us to think that what we call 'the body' or 'our body' is simply something that we have decided to leave unconceptualized, precisely to give ourselves an illusion of *Grund*, of foundation, where there is in fact only the indeterminate, imaginary effect of what is determined as symbolic, which we merely reflect as if in a mirror. Why retain the name of *ethics* for acts relating to a so-called 'body' including all technical activities, as if they only waited to get their meaning from it? Is it not rather what we called 'my body' that should be treated as problematic, and which we should learn to do without, basing ourselves on the development of those techniques which are always changing the linkages and networks we are forming with machines, without reaching any destination or end? We do not in any way think that the images that medicine has supplied to us, of our cells, tumors and fractures, alienate us, as long as we refrain from relating them to a body, about which we seem sometimes to regret the good old times when it was touched and palpated by the doctor, as if that represented a timeless and priceless advantage. We do not see why the fact of being touched and palpated should give the body a more 'real' status than that of being the imaginary of the regret for old-fashioned medicine. That images are assumed in the mode of a re-appropriation in the name of 'my body' is admittedly no more questionable than other attitudes, but that concession does nothing to substantiate any claim on behalf of the body to a monopoly of truth, or to be regarded as the necessary foundation of all treatments. There are many other ways of appropriating a medical examination or treatment besides believing that one experiences it upon the foundation of 'one's body', and we cannot see what ethics has to gain from singing that old song of Merleau-Ponty's.

It is the same with telemedicine, which worries some of our contemporaries for similar reasons, to the point that some see, in the image of the unfortunate moment when a body will be digitized, filed and fixed, the foreshadowing of a grave threat of the violation of the autonomy of the patient.[20] The image is no truer when it is supposedly related to 'my body' than when the body is treated in the symbolical way that we have just mentioned. To be offended or concerned by that treatment, one must think of the symbolic as an alienation of a body supposedly lived and deemed more authentic, or which should be lived more authentically. However, there is no requirement for anyone to think things are actually that way except the defensive, reactive and preservative attachment to a sort of stability of the human subject or to a

[20]Echoing that unfortunate corporeity, Professor Didier Sicard wrote, in a book which is quite remarkable for its prudence and the information it contains, ([45], pp. 163–164): 'Ce corps, déjà mis à distance par l'imagerie diagnostique, va désormais 'être envoyé' sur les écrans de la télé-médecine. Numérisée, archivée, fixée, transmise à l'autre bout du monde, cette image d'un moment malheureux du corps devient une information organique. Le statut de l'autonomie de cette image est un des problèmes du futur. Pourra-t-on retravailler cette image, la virtualiser? ou n'envoyer que des images définitivement fixées? On imagine les problèmes éthiques que pose l'intervention médicale sur une image . . . qui pourrait ainsi être rendue artificiellement pathologique ou normalisée. . . La télé-médecine pose comme problème majeur celui de l'autonomie du patient face à la médecine.'

human nature – the existence and stability of which are as impossible to demonstrate as the belief that ethics can only consist in such an upholding, desperate though an activity which one knows in advance cannot succeed may appear to be. Lacan was right to note that the *body* of Merleau-Ponty's or Goldstein's phenomenology only duplicated and inverted all the functions of the *mind*. 'Ce n'est pas pour autant, ajoute-t-il, que nous devons nous trouver satisfaits, car il y a tout de même bien là quelque escamotage', when, 'après que de longs siècles nous avaient donné dans l'âme un corps spiritualisé, la phénoménologie contemporaine nous fait du corps une âme corporéisée.' ([24], pp. 253–4)[21] The word *escamotage* is quite well chosen insofar as the reference to the 'body' precisely masks the emancipation of machines (that are being coupled with the body) from any imitative servility.

6.5 Fine Words: 'The Fusion with the Machine Is Always to the Detriment of the Human'

Rather than being content with big ambivalent words pronounced with hand on heart, or with the tragic expression of a righter of wrongs,[22] it would be better to pose the problem mathematicians call inverse: what, in the compound that the human being constitutes with the machine(s), could either prevent or obstruct the trace or the graph that we are talking about? Or, if we agree that that some trace always happens, what could render it unacceptable, inadmissible, or dangerous, and what makes one source of trace so dominant over all the others, so exclusive, that it could not coexist with them, and would not be capable of combining with them in any dialectics or history? One should, in relation to the generation of new machines, continue the patient work that Simondon started on the mode of existence of the technical objects he knew, without inflecting ethical questions from the outset with a moralism that an insufficient ontology renders obsolete, and which does not leave any room for anything but emotion. One knows all too well the answer transhumanists give to this type of

[21]In the article he wrote on Merleau-Ponty upon his death, Lacan, showing his difference with to latter, wrote that 'si le signifiant est exigé comme syntaxe d'avant le sujet pour l'avénement de ce sujet, non pas seulement en tant qu'il parle, mais en ce qu'il dit, des effets sont possibles de métaphore et de métonymie non seulement sans ce sujet, mais sa présence même s'y constituant du signifiant plus que du corps, comme après tout on pourrait dire qu'elle fait dans le discours de Maurice Merleau-Ponty lui-même, et littéralement.' Lacan however, performed a generous reading of his friend who died prematurely. It seems that it is rather, and most often, in opposition to the phenomonology of the body [28, 29] that Lacan asserted that 'le corps dont il s'agit en psychanalyse est le corps d'un être qui parle et de ce fait même, voit son fonctionnement organique profondément altéré et transformé par cette incidence du langage.' The body is changed by the fact that it is engaged 'dans la dialectique signifiante'.

[22]We know the statements: 'I would rather live until the age of seventy or eighty than live for two hundred or three hundred years,' 'What would our relationship with others become if we lived that old?', and 'Would we not be an unbearable weight for society?' Or perhaps is it a very odd debate, particular to some French ideologists!

proposition: you give too much prominence to symbolism when you believe in the virtue of an antinomy one thesis of which would be occupied by robotics and nanotechnologies. Symbolism does not ensnare robotics anymore, but it is rather robotics and new technologies which clamber aboard <*arraisonnent*> the symbolic and condemn it to inflation, to borrow two famous expressions, one from Heidegger's translators, and the other from Gilbert Hottois [17]. In the real silence and mutism of the end of language, even though empirically speaking and apparently language is proliferating, technologies win the game they play, without sentences, with the symbolic, in a sort of meta-dialectics which does not have much in common with that which Kant left us [22]. What would be at stake in the meta-dialectics created by the opposition of technique and symbolism would not be another antinomy which would have been forgotten by Kant, and which would oppose to the thesis according to which 'les machines les plus compliquées ne sont faites qu'avec des paroles',[23] the antithesis according to which 'le monde symbolique, c'est le monde de la machine',[24] but an antinomy that would be for the most part phantasmic, which would transcend the other antinomies and would doom dialectics itself, and with it language and history, to extinction.

It is no wonder that Kantian circles are so interested in the issues of transhumanism, for despite the amazement provoked by the overcoming of dialectics by new technologies, those issues are exactly where Kant located the limits of our knowledge: death, the other (the other man or individual), what things are in themselves, the relation to time which renders us as far removed, because of our confinement in the present, from the past as from the future. The linkage of the 'human' individual with machines, which shakes his nature and his boundaries to the point of compromising them, compels thought on these strategic points. Once again this should not surprise us. However, we may find it more surprising that they are sometimes so mishandled by precisely that transcendental realism from which Kantian dialectics seemed to have delivered us [22].

6.6 Transhumanism and Transcendental Dialectics

We are told loudly that, potentially within a few decades, we are going to have the possibility to live for several hundred years. The ignorant, who do not know it, do not attempt to measure the significance of such news, yet are made fun of. We should even, supposedly, attempt to bridge the gap that lies between us and the thinkers of

[23]Which may imply that there must be speaking beings to build machines.

[24]This sentence is echoed in a passage of the Conference delivered by Lacan on 22nd June 1955 ([23], p. 350): 'On sait bien qu'elle ne pense pas cette machine. C'est nous qui l'avons faite, et elle pense ce qu'on lui a dit de penser'. Lacan however does not say this to take a Cartesian path, for he at once adds that 'si la machine ne pense pas, il est clair que nous ne pensons pas non plus au moment où nous faisons une opération. Nous suivons exactement les mêmes mécanismes que la machine'.

transhumanism. What do these prophets, who order us to say that that long life is good – while others think it is bad, and thus ask us to take their side – know more than us? They, of course, do not know more than us about death, and are only making themselves ridiculous by declaring, with apparent seriousness, that they would rather die than live for two more centuries or, on the contrary, that they would not hesitate to add two more centuries to our current paltry life expectancy. That supposed multi-century existence, which one welcomes or rejects, is as unthinkable as death. We are not rejecting that hypothetical claim, but we do not know how to identify that imaginary 'phenomenon', we do not know to whom it will happen, upon which conditions it will happen, and, to be honest, we are not that interested in it; no more, in any case, according to Hume, than we are interested in the supposed life that some would have us believe we may enjoy after death.[25] The two or three hundred years that we thus grant ourselves – thanks to what claims to be much more than a thought experiment – are the attempt to fill a gap linked to an unrestrained use of language[26] by imagination, and which perhaps one should not be encouraged to fill in too realistic a fashion. This gap is also not that of the projection of an increase of our faculties in the sense they seem to indicate – see more, hear more, go faster, etc. – but that of incredible invisible interstices between these augmented faculties. These gaps do not prevent the development of sciences, nor any practical activity, and do not at all require to be filled by the type of phantasm which becomes ridiculous when it takes on grandiose pedagogical or inspired airs, instead of adopting the modestly critical tone that better suits philosophy.

We project ourselves out of our present towards the past or the future with the same 'success' as a long time ago. This projection only has meaning if it allows for some consistent work of construction. This 'we' has never had any guarantee of stability except that with which it is endowed by guile and fiction. This flimsiness of this guarantee by no means delegitimizes any and all work that is stretched and extended from the information one has at one's disposal towards the past as well as towards the future, while recognizing that the subject who makes the assessment is always in a difficult situation in relation to the positions he aims at in a time in which he is not, never was, and never will be. Is it tragic that we do not know how to fill those gaps of the past and future? Does our ignorance stop us from trying to, or having to, act? We do not easily give that question a Kantian meaning. Utilitarian calculations give themselves the possibility of overcoming time distensions and the intersubjective distances that exist, despite all our knowledge, between one time and another, one subject and another, one individual and another. How could it be different to live in the present, since we never have the guarantee that we are and will be the same beings as men from the past or the beings that will be to come?

[25]'For I observe, that men are everywhere concerned about what may happen after their death, provided it regard this world; and that they are few to whom their name, their family, their friends, and their country are in any period of time entirely indifferent.' ([19], p. 79).

[26]One can, therefore, when one is in Paris, wonder what time it is on a given point of the sun or of some other star: this question is at the same time unavoidable and absurd. It is due to a lack of control of the limits of language.

Transhumanist discourses are strange undertakings which try to give some consistency to a real thing that does not and cannot have any. They try to replace the uncomfortable non-knowledge of a dialectics by filling the voids with the products of a supposed knowledge which we were taught, and constantly have to be taught, to do without. The height of their logical absurdity is, after scolding the reader for not taking sufficiently into account the horrors that await us on all sides – extinction of the species, monstrous cyborgs, policies necessarily considered as totalitarian – to revert to a perfectibilism that, in the end, we will have the possibility to choose, like those which one can find in Rousseau or Kant. Was all that fuss really worth it? All the more so since, if a page has really been turned at this time of 'intelligent machines' and robots, it is really that which closes on the conception of a democracy as contemplated in the way of Rousseau and the contractualists. We would like to sketch this topic, which is more political than ethical, in more detail before ending this chapter.

6.7 Reflections on What Democracy Becomes at the Time of Information Technologies: A More Political Than Ethical Parenthesis[27]

Rousseau wrote in *The Social Contract*: 'If there were a people of gods, it would be democratically governed. A government as perfect as that is not suitable for men.' ([40], p. 406; [41]). That is a way of saying that if men were capable of forming a perfect government, they would not need government – as Rousseau had highlighted some lines before: 'A people who would always govern well would not need to be governed.' ([40], p. 404). Would such a government be more suitable for beings that some hesitate to call 'human' because of their linkage with machines and technical nets? Would it be suitable for beings whose capacities are increased by technical means? Or, on the contrary, would not the situation worsen and, as in the title of an English book translated in French, would the Net not make people so dull that it would lead us far away from democracy, even though its founders have always claimed – sincerely or hypocritically – that their invention was democratic because it liberated individuals from the power of journalists, of scholars teaching in universities, and of the authorized representatives that have made their careers in many political bodies? Would it not facilitate the accretion of still more wealth and more scandalous power in fewer and fewer hands, as Yuval Noah Harari,[28] the author of *Homo Deus* [16], has suggested in what one may call a prophetic and visionary

[27]We have already explained what we thought of the great difficulty of containing ethics in very strict limits, given its 'circumstantial' character that we have underlined in our first volume and in the foreword of this volume: see [10], pp. 98, 148; and p. VIII above.

[28]See Shapin S., ([44], pp. 30–35). The book the article refers to was published by Cambridge University Press, 2014. P.-E. Dauzat translated it into French with the subtitle: *Une brève histoire*

style? Is it not striking that this narrow concentration is equally incompatible with both the democratic ideal and its reality that consider everybody as equally eligible for all public dignities, places and employments, without distinction on any other basis than virtues and talents? Would not the development of these technologies, or at least of some of them, prove fatal to democracy?

This incompatibility could even go further and strike a blow not only at democracy but also at government itself. In a gripping passage in *Phaedrus* [39], Plato – who was not a democrat – presents, in an Egyptian story, as he calls it himself, a heated exchange between a god, who is proud of his inventions and shows off samples of them, and Thamus, the king of Thebes, to whom he extols them. The god points out to the king that he has invented numbers, arithmetic, geometry, astronomy, draughts and dice, and, most important of all, letters (γράμματα).[29] After this presentation, the god announces to the king that these inventions ought to be imparted to the other Egyptians. The debate is fierce – which is astonishing for us, when we reflect that *Epinomis* presented the invention of number and calculation, even more than language, as the special feature of human nature from its origins – and the king is very critical of the uses of these various techniques. But it is when he presents himself as a representative of writing that, although he is a god, Theuth is given a lesson by the politician who has other and wider responsibilities: 'Most ingenious Theuth, one man has the ability to beget arts, but the ability to judge of their usefulness or harmfulness to their users belongs to another' (274 e). The king continues with an illustration of the damage caused by writing, and the seeds of the destruction of human nature that it carries within itself. The king's words are left unanswered, or at least we are not told what the god replied. One can imagine modern Theuths presenting the merits of computers to those modern Thamuses who have become presidents of democracies or of republics: what might be the reply of these contemporaneous Thamuses to the Theuth delegated to meet them on behalf of Silicon Valley? Nowadays, we recognize some Thamuses, whose language is very different from the conservative language of their Egyptian ancestor, and who are ready to encourage the expansion of informatics and of artificial intelligence which might bring about a change in humanity perhaps as radical as the changes brought about by letters in his time. We will try to consider whether they are right or wrong, and avoid dealing with these questions by mixing them with mythical stories that are, at present, not Egyptian but, for instance, posthumanist. This is to say, we will try to be as critical as possible and steer clear of the reveries of power and myths of 'enhanced humanity' with which we have been inundated in certain recent works – some of them written by Kantians.

---

de l'avenir in 2017. S. Shapin's article, translated by D. Veaudor had already been published in the *London Review of Books* on 13th July 2017.

[29]It is quite strange to note that the inventions and techniques presented by the god Theuth are among the main functions expected of present-day computers: games, learning languages, mathematics, calculation, and memorization, which is at the core of *Phaedrus*.

1. During the last three or four centuries during which democracy has been taken seriously, it has almost always been thought about in terms of ideal patterns that isolate the individual who is held to be the fundamental political unit, leaving aside the features that make him a real individual. The point is particularly striking in Rousseau and Montesquieu. Rousseau does not satisfy himself with calling democracy a system, alongside monarchy and aristocracy. Like Hobbes, he thinks, basically, that the contract that legitimates any government is a democratic contract. Everybody, making his own decisions on a problem that he understands well, thinks about the social body and decides for it what he thinks is best. A contract is the relation of the whole people to itself through the mediation of the individuals, which results in everybody choosing for the others when choosing for himself, while the others, when choosing for themselves, also choose for one. The individual in *The Social Contract* does exist, but remains in the background. This is how a structure called *la volonté générale*, *the general will* – which he distinguishes from the *la volonté de tous*, *the will of all*, which is nothing but a disordered collection of the particular wills which confront each other without any common measure – is supposedly built. Such a common denominator would shape the matter of the laws that citizens need to regulate social life or individual life in a social community. Laws are the conditions of the existence[30] of the general will, and even its acts.[31]

Rousseau could not completely overlook the fact that he had drawn a simple and abstract sketch of democracy, which was so ideal, as Montesquieu had seen in the chapter on democracy in *The Spirit of the Laws* [32], that, when it was deprived of the *principle of virtue* that must lead every citizen, even the young ones, it was dangerously out of order. Surprisingly, Rousseau, who was more radical, did not hesitate to say that democracy had never existed, did not exist and would never exist. It would imply – and so does the social contract[32] – that everybody, against the

[30]*The Social Contract*, II, VI: 'Les lois ne sont proprement que les conditions de l'association civile.' ([40], p. 380).

[31]'Les actes du souverain ne peuvent être que des actes de volonté générale, des lois' ([42], p. 842). There is an identification of the general will and the law: in a note to Book II of *The Social Contract*, Ch. VI, defining what he calls a republic, Rousseau says that it is told 'en général, (de) tout gouvernement guidé par la volonté générale, qui est la loi.' ([40], p. 380). <in general, any government directed by the general will, which is the law (trans. J. Bennett, earlymoderntexts.com/assets/pdfs/rousseau1762.pdf)>.

[32]The point is made in a remarkable passage in *The Social Contract*, B. III, Ch. IV: 'Voilà pourquoi un auteur célèbre a donné la vertu pour principe de la République; car toutes ces conditions ne sauraient subsister sans la vertu: mais faute d'avoir fait les distinctions nécessaires, ce beau génie a manqué souvent de justesse, quelquefois de clarté, et n'a pas vu que l'autorité souveraine étant partout la même, le même principe doit avoir lieu dans tout État bien constitué, plus ou moins, il est vrai, selon la forme du gouvernement.' ([40], p. 405). <That is why a famous writer – Montesquieu – has made virtue the driving force of a republic; for none of the conditions could exist without virtue. But that great thinker did not make all the needed distinctions, and that lead it often to be inexact and sometimes to be obscure; he did not see that because the foreign authority is everywhere the same, the same driving force should be at work in every well-considered state -more or less, it is true,

natural trend, prefers the general interest to his own, and that he prefers it continuously, throughout his life. This is not possible. To this moral argument, which implies that, in a democracy, everybody must be a hero or a god, some arguments, which are as essential but which the classics had not especially heeded, must be added.

Before expounding them, we must underline a paradox. Rousseau and other contractualists drew up precise chimerical contracts and abstract sketches of democracies whilst simultaneously admitting that they could not assess men's actions, which results in their citizens being no more than phantasms of purity, and their democracies no more than phantasms of unaminity and accuracy derived from the application of only crude measuring instruments. They might have been better advised to satisfy themselves with good approximations of measures that can really be applied to the real actions of real men. Of course, these contractualists – and Rousseau in particular – often protested that they took men as they were and the laws as they might be; but the reality is that, in order to establish a contract which is supposed to measure and balance citizenship in the most fundamental way, they only ever took into account a relationship of understanding and will, completely neglecting the concrete and emotional life of men without whom, however, no such contract can be fulfilled, or even make sense.

2. What decisively prevents the Rousseauist or contractualist democratic ideal from being useful is that political and social action involves more than a set of beings that are reduced to understanding and willing agents likely to contract with one another. The evanescence of these individuals is such that they are required to vanish in the relation in which the social body stands with itself.[33] The limits that delineate the individuals and separate one from another are not given. It is very difficult to develop a defensible conception of political relations if one begins with individuals whose minds are somehow hermetically sealed one from another, even when one assumes them capable of focusing simultaneously on themselves and the social whole by some kind of science. Without dealing here with the question of why it would be better to solve political problems by asking individual citizens in isolation to consider for themselves and consciously what is

depending of the form of the government (trans. J. Bennett, earlymoderntexts.com/assets/pdfs/rousseau1762.pdf)>. [41]

[33] *The Social Contract*, II, VI: 'Quand tout le peuple statue sur tout le peuple, il ne considère que lui-même, et s'il se forme alors un rapport, c'est de l'objet entier sous un point de vue à l'objet entier sous un autre point de vue, sans aucune division du tout. Alors la matière sur laquelle on statue est générale comme la volonté qui statue. C'est cet acte que l'on appelle une loi.' ([40], p. 379). The Geneva manuscript went as far as saying (II, IV): 'La matière et la forme des lois sont ce qui constitue leur nature; la forme est dans l'autorité qui statue, la matière est dans la chose statuée' (Pléiade, p. 1461). <When the whole people decrees for the whole people, it is considering only itself; and if a relation is then formed, it is between two aspects of the entire object, without there being any division of the whole. In that case, the matter about which the decree is made is, like the decreeing will, general. This is what I call a law. (Trans. J. Bennett, earlymoderntexts.com/assets/pdfs/rousseau1762.pdf)>. [41]

for the common good of the community, we must first ask whether it is possible. Even supposing that political decisions could be made by this method, how would it be possible for anyone to recognize them, and differentiate them from illegitimate and corrupt decisions? The ideology of deciding with reference to the condition of the others and on behalf of the others only, based on one's ideas of the common interest, neglects from the very beginning the fact that there is no individual mind that can be absolutely independent of the others. No mind can address itself without that address implying the existence of other minds, which imply its own existence in their parallel self-reflection. Nobody knows the furthermost bonds of a mind nor of the debt owed to the others in the recollection of oneself. These limits cannot be decided. The idea of *alienation* – whether we mean *Entfremdung* or *Entäusserung* – implies a dogmatic idea of the individual. It would be impossible to put on trial, in the name of a chimerical autonomy, a heteronomy which is a necessary component of the minds of men whose vocation is to change and develop continuously. This 'becoming other', thanks to books, encounters, passions, travels, cares, work and the uses of instruments and machines, does not always contain the loss or lessening included in the idea of *Entfremdung*.

Men's minds can never develop but by integrating otherness, and it is simply impossible to isolate a self completely. Integrating otherness includes mixing with a technical world that flows through us as a flux in which we cannot, in spite of our illusions, claim to be subjects confronted by objects that could be machines or any other technical objects. The technique is fundamentally different from the technical object and is not any aggregate of such objects, but is a sundry web of practices, uses, consumptions, creations, and innovations. In consequence, any subject who supposes himself confronted with a multitude of objects, and who not only thinks that he can realize an eidetic reduction of himself, but goes on to insist on making such a reduction the indispensable starting point for political action, does no more than lose himself in the most perfect illusion. We have been participating in the world of machines – whether mechanical, chemical, or digital (with artificial intelligence) – for a long time. Our lives multifariously depend on these machines, without it being possible to separate the former from the latter. They are composed of them, as our organs are composed of cells, and as the ideas we have come from beings who do not know their limits and who are inextricably connected to technical nets. We should not be disturbed by this thought, but it is necessary to understand how humanity – if it is still possible to use that term – is inserted in those nets, particularly when the machines with which it is coupled begin to perform on its behalf a number of tasks that it used to believe could be carried out only by human judgment and by some other intellectual functions by which it supposed it was identified.

3. Intelligent machines – if might dare use the expression – function in the same way that the mind, which constitutes them, and which has itself always been already constituted by them (until, one day, machines constitute themselves), can turn and act upon itself, and elaborate a new level of itself, as in mathematics, when, in order to solve a problem, it builds a symbolic level by means of which it goes

> beyond its own intuitions and can think without representations. The 'human' mind can think by encompassing and folding itself in symbolic levels, which are increasingly more complicated, not representable, and which will in their turn become intuitive for other foldings and envelopments.[34] We could not say whether the process of development and growth through increasingly complicated levels is deeply human and we do not care. It gives a good account of the game the mind plays with itself, of the game of the complex systems to which it contributes, without it being possible to say whether there exists any level where the mind is itself, or whether the mind is between levels of innovations that push it into a constant state of disquiet and movement. We do not have to settle at some supposedly human level, from which we should judge that some others are not human. Ethics itself, with the use of which we would like to establish such a level, works in the same way, as R.M. Hare has shown in *Moral Thinking*. Pascal, who invented calculating machines, fell into the trap of attributing some truths to the heart, which he then had to overcome when solving problems.

As soon as men began to make machines work efficiently – by which they overcame the failings or limits of their own faculties and substituted for their own intelligence – they immediately understood that their links with the technical world and the bonds between them could never be the same again; that humanity which was until then their context of reference had become uncertain in its outlines and limits and was disturbed in the self-certainty it had just won. Politics itself was changed, and those inventors of machines, who generally challenged the idea of the *self*,[35] would never have thought of promoting some social contract, the laws of which would be strictly deducible from any system depending on a conception of self-determining, understanding and willing subjects, who co-operate as equally free individuals, so that they warrant equality by liberty and liberty by equality.

This conception of the law is idealistic and radically incompatible with another other iron law according to which man changes by constituting increasingly complicated symbolic levels. It claims to promote a *volonté générale <general will>* which in no way coincides with the bundle of singular wills which do not take the others into account at all. However, we only ever have at our disposal the *volonté de tous <will of all>*, from which we extract the *volonté générale* which presents itself as an *a priori*, although it can only ever be a result. The general will is nothing substantial, and it is necessary to construe it, if any sense can be made of that. The illusion of Rousseau was to believe that the law could be derived from the *volonté générale*, although it is from the writing of the law that we must start to build – if in this relation any common measure is possible – what articulates the laws, and doing so will remain a thorny problem. Rousseau began with the end. He did what early

[34]It is interesting, in this respect, to see that that notion of *folding* also works for authors as diverse as Spinoza (*involvere* in latin), Jaspers (*das Umgreifendes* in German) and Deleuze, who all reject the concept of foundation.

[35]As is the case, for example, of Pascal. ([34], p. 1135), ([35], pp. 217–8).

mathematicians did when, in their analyses, they held the problem to have been solved and thought it possible to infer the solution in that way.

Now the problem is rather the reverse: the issue is less to infer the law from the general will than to attempt to define the laws, to adjust them to human needs and to articulate them without deducing them. Who would be content, in order to solve some problem in physics, to satisfy themselves with saying that it is nature which connects laws together? Just as the work of the physicist consists, not in understanding the phantasm of a nature, but in calculating the interference of physical laws one with the other, without which nature would be a mere fiction, so the work of the lawyer lies in forming and articulating laws, and not in the mere phantasm of first denying and then plucking from the ether the basis of the social bond.

4. However, if laws can have neither the form nor the matter ascribed to them by Rousseau and the other contractualists, who seem to hold the key to the ideological principle of democracy, then what form and which matter is it possible to give them in order that centres of thinking, calculating and deciding, which are inextricably coupled with a world of machines – particularly of intelligent machines – may regularly coexist, while taking into account the circumstantial complexities of their coexistence?

The pattern of law that seems best equipped to solve our problem can be found in part in the definition given in Bentham's *Of the Limits of the Penal Branch of Jurisprudence* [5]. Sometimes its apparent difficulty and complexity makes student-lawyers throw up their hands in despair. 'A law may be defined as an assemblage of signs *declarative* of a *volition* conceived or adopted by the *sovereign* in a state, concerning the conduct to be observed in a certain *case* by a certain person or class of persons, who in the case in question are, or are supposed to be, subject to his power: such volition trusting for its accomplishment to the expectation of certain events which it is intended such declaration should upon occasion be a means of bringing to pass, and the prospect of which it is intended should act as a motive upon those whose conduct is in question.' ([4], p. 1; [5], p. 24).

Every law is a text that gathers, at least for a time, majority support, and might disappear when it loses that support. So the plots and deals <in French: *brigues*> which Rousseau feared so much are not necessarily harmful to democracy. Far from being impediments to political expression, intrigues – as Montesquieu calls them, using the same word <*brigues*> as Rousseau – are essential components of it.[36] The

[36]In B. II, Chap. II of *The Spirit of the Laws*, Montesquieu says, startling a reader familiar with Rousseau, that 'la brigue n'est pas dangereuse dans le peuple, dont la nature est d'agir par passion. (…) Le malheur d'une république, c'est lorsqu'il n'y a plus de brigues; et cela arrive lorsqu'on a corrompu le peuple à prix d'argent (…)' ([31], p. 244). <Intrigue is dangerous in a senate; it is dangerous in a body of nobles; it is not dangerous in the people, whose nature is to act from passion. (…) The misfortune of a republic is to be without intrigue, and this happens when the people have been corrupted by silver> ([32], p. 14); <Intriguing [though dangerous in a senate, and also in a body of nobles] is not dangerous in the people, whose nature is to act through passion. (…) The misfortune of a republic is, when intrigues are at an end; what happens when the people are gained

system of laws – since every law must intersect in a co-ordinated pattern of articulation with many others – is a sort of construction of commands temporarily supported by a majority of its subjects. The articulation of the laws is also a problem that demands solution, but never definitively and permanently, because laws change. It does not require any such strong principle of unity as is viewed as its essential condition in Rousseau. The utilitarian conception recognizes an almost casual aspect of the law, which is circumstantial and contingent, whereas Rousseau conceived law as strictly rigid and formal.[37]

Do not think, however, that utilitarianism is the only theory to defend such a conception of the law. Oddly enough, this utilitarian conception is closely akin to Ricoeur's thought. Of course, Ricoeur would not have objected to the discourse of conscience, but he saw in the law less a consensus bound to the *volonté générale* which is supposed to be at the foundation of the State than, in a divided world in which there is no hope of reconciling all its members, a warrant for any member to play its game, knowing that divisions will be arbitrated rather than reduced. Democracy requires no more than honestly taking account of the majority which is associated with every law, and fairly considering how every such majority comes to an agreement with another.

This conception of the law fits best with the sort of democracy that we are looking for, and which does not require a common measure which is given at the beginning and *a priori*, but which must be created. For it is neither stated nor required that the majority supporting every law should be always the same. It varies from one law to another, always creating residues, but not always the same ones. Thus, this legal play of majority suits a certain style of democracy which does not look like the regime promoted by Rousseau. 'The only species of government which has or can have for its object and effect the greatest happiness for the greatest number, is a democracy' ([2], p. 47). But this 'greatest number' implies neither that it be made up of individualized beings, nor that it always be the same individuals who give their assent to every law, and in the same way. After listing the characteristics of a pure

by bribery and corruption. (trans. in *The Complete Works of M. de Montesquieu*, London, T. Evans, 1777, 4 vol., vol. 1, all.libertyfund.org/totles/montesquieu-complete-works-vol-1-the-sprit-of-laws [32]). On the contrary, Rousseau condemned plots without exception in *The Social Contract*, B. II, Chap. III: 'Mais quand il se fait des brigues, des associations partielles aux dépends de la grande, la volonté de chacune de ces associations devient générale par rapport à ses membres, et particulière par rapport à l'État; on peut dire alors qu'il n'y a plus autant de votants que d'hommes, mais seulement autant que d'associations. Les différences deviennent moins nombreuses et donnent un résultat moins général'. ([40], pp. 371–2). <But when plots and deals lead to the formation of partial associations at the expense of the big association, the will of each of these associations – the general will of its members – is still a particular will so far as the state is concerned; so that it can then be said that *as many votes as there as men* is replaced by *as many votes as there are associations*. The particular wills become less numerous and give a less general result. And when one of these associations is so great as to prevail over the rest, the result is no longer a sum of small wills but a single particular will; and then there is no longer a general will, and the opinion that prevails is purely particular. (Trans. J. Bennett, earlymoderntexts.com/assets/pdfs/rousseau1762.pdf)>. [41]

[37]The matter of the laws being deduced, in a way, from their form.

democracy in which 'all is regularity, tranquillity, prosperity, security, continual security, and with it continually increasing, though with practical equality divided, opulence' – he could have added liberty, equality before the law, and perhaps economic equality, which are notions at the core of contractualism – Bentham added: 'As for us, we need no such purity'; 'not pure democracy do we want, nor anything like it: what we want is, under the existing forms of subjection, the ascendency –the virtual and effective ascendency– of the democratic interest: this is all we are absolutely in need of: with this we should be content: with less of this it is in vain to speak of content: for less than this cannot save us.' ([3], p. 47).

Of course, this conception of democracy and of the law cannot solve all the problems. The issue of ethnic or national minorities remains unsolved by the realistic but dangerous famous phrase of utilitarianism – the greatest happiness of the greatest number. How might minorities be respected and protected?[38] By what processes is it possible to get rid of laws that are no longer relevant for a political community, and that are progressively rejected through evidence of public discontent?

There is no reason to presume that the extraordinary variety of possible relations to the law, added to the huge diversity of the relations that citizens may form with each other, will corroborate or seamlessly accommodate the extraordinary variety of modes of creation, use, acceptance and refusal of digital technologies. Our claim is rather that the prodigious and continuously developing variety of the digital gives rise – or at least is capable of giving rise – to always new and always more refined laws which might attract always renewed majorities and minorities. Simply to presume the inevitability of the contrary, to conclude without evidence that digital culture is intrinsically opposed to certain modes of political organization, democracy among them, and must poison them by perverting traditional typologies of government and introducing confusion within their limits, is not thought, but precisely the abdication of the responsibility to think seriously.

5. Would there be, in the digital culture, in the working of the Internet, a principle – we borrow the term *principle* from Montesquieu, who distinguished the *nature of a regime* and its *principle* ([31], pp. 250–1) ([32], p. 21) – which would radically prevent it from functioning democratically? Surely our democracies are far removed from the ideal of virtue that Montesquieu imagined to be necessarily their principle ([30], pp. 250–1) ([31], pp. 22–4). At best, our democracies operate without virtue, and they nearly succeed in doing so, except for the most scandalous inequalities. However, does not the absence of any virtue, and the presence of cynicism, outbursts of hatred and of anger, shameless and deliberate lies, deliberate and arrogant ignorance, the avoidance of any debate with those who are more learned, the disdain for any instruction, the cowardice that disparages others unjustly and hides behind anonymity, not reach a peak through this medium? Is it not paradoxical to conceive machines and networks that require, for their conception and development, very sophisticated symbolic levels in

[38]It has often been said that one can recognize a democracy not but by how the law of the majority is respected, but by looking at the fate of the minorities.

mathematics, which eventually facilitate the effusion of a tidal-wave of public and private words – words the plethora of which makes them cheap, desecrated, or trivial, even when they are not poisonous and harmful – which flood the fields of politics and religion? Has there ever been so much imagination displayed over an issue that seemed to promise democratically to call on everybody to speak, without distinction of rank, job, birth or competence, but that ends up dominated by ill-informed if not self-consciously duplicitous, speech and writings? It has become unusual and seems to be considered a waste of time to go back over what has been said or written or to scrutinize it in any detail, and speech is carried away in algorithmically amplified whirls of noise that undermine it completely.

It is indisputable that the fear of shame, without which no civilized society is possible,[39] is no longer a powerful motive among those who make no secret of their political views on the Net, regardless of all truth and fairness. In regard to the innumerable testimonies that are poured on the Net, shame no longer acts as a check to prevent falsity. One may express hateful or venomous words about persons that are littered with allegations of offences they have not committed, without being forced to drop one's mask and declare one's identity, thus making it possible for others to repeat them, unscrupulously and in a completely hypocritical way, as if they trusted the sources of information. This generalized regime of anonymous letters, of which the addressee is often the last to know, removes from us a virtue to which we are no longer obliged to pay respect, and without which there is no longer a State – at least no longer a democratic state. The risk of losing face while writing on the Internet is nothing like as high as when one makes a speech, writes an article in a newspaper, writes a book, participates in a radio programme, or appears on television.

In the name of *freedom of thought*, one thrusts forward one's opinion without bothering about the quality of the argument or evidence, and without wondering whether one injures somebody. One becomes used to considering such excesses as so much collateral damage of what is still called *freedom of thought*, as if this freedom did not involve any duty, and as if, stripped of all obligations as it is, it were not sometimes a sort of violence for which individuals, persons and also collectivities and the greater community have to pay. Sentences which are deprived of any critical thought and which circumvent judgment, offend reason and even simple common sense. Before any exercise of judgment, gross errors are permitted to be established alongside the products of critical intelligence –including the most refined. Insult and invective take the place of symbolic discourse, and claim an equal or superior status.

---

[39]Bentham identified it as an essential political passion in *Rationale of Judicial Evidence*, ([1], I, p. 422): 'Shame may be considered as operating in the character of a security for trustworthiness in testimony, in so far as, on the occasion of a man's delivering testimony, the contempt or ill-will of any person or persons is understood to attach, or apprehended as being about to attach, upon a deviation, on his part, from the line of the truth.'

But if the display of all these knaveries has perhaps never been so obvious as nowadays, were there not, in previous centuries, newspapers that allowed nearly the same thing? Was everything disseminated on the radio or through phone calls of better quality? Would the ballot that some voters placed in the box, after stepping inside the polling booth, not have made them feel ashamed if it had come to be known by all?

Is it true that it is easier, for individuals, parties, political or religious communities, to lie on the Internet than to lie in newspaper articles, books, or through cinematographic or televisual pictures? Is one less on one's guard and more easily influenced when one is bathing in the flux of the Internet, even when one is adequately educated, than yesterday when one was watching films or television, or reading articles or books that disseminated some ideology or propaganda? Why should it be easier to lead somebody to adopt positions that he would receive 'from outside' as if they were his own? Would being flooded with information make one more vulnerable than being deprived of it, and getting only a few pieces of information or news? The selection of the items of news which deserve attention – whatever the medium through which and the time at which news is received – always remains a difficult task for the addressee, from which he cannot escape if he wants to be adequately informed – so that he may explain his votes and decisions. We have always received the major part of our information through testimonies of others, and we can neither check it piece by piece, simply because it is impossible, nor entrust some State organ to screen all pieces of information, establish which ones are right and which ones are wrong, because this task cannot be carried through to a successful conclusion without lapsing into authoritarianism. Certainly it is better to have an Internet carrying all sorts of fallacies and falsities. We can understand it all the better because nothing could prevent the police from exploiting an inexhaustible source of documents.

When the Internet is indicted for allowing foreign pressures on the ballots of the citizens of a State at the time of an important election, is it not possible to answer that there have been, at all times, mass and collective mistakes? Were there never people who were mistaken, though they imagined they were right, and people who deceived other people more or less knowingly, and sometimes brazenly? Does history not report rumours and panics? Are not these phenomena sometimes so many ancestors of fake news? Does it not always require, at any time, the greatest care and serious effort to go against the trend, and how is it be possible to argue that these efforts are called for against a danger and difficulties that are specific to our time?

The two biggest worries for democracy in our digital age seem to be as follows. The *first* is that it is possible, with these technical means, to give one's advice and to avoid the debate which is the very essence of democracy, and therefore to kill democracy while pretending to participate in it. The *second* consists in the constitution of big databases which result in the fact that States, which must protect their citizens, are mainly discredited in their sovereign task by the giants of the Internet (Google, Apple, Facebook, Amazon) which accumulate and store much more information than is possible for States to do. Moreover, it is not enough to store data. It is necessary to use the data wisely, or at least responsibly. There are technical devices to refine control. These huge memories, which know more things about each and every individual than he may know himself, because he has had time to forget what he used to know, these masses of information which make it possible to know what anyone is doing at any moment, are dangerous for two reasons that threaten to trap us. Either States can use them, but they depend for this use on big companies that could refuse, or demand that the user pay for the service; or the states are bypassed in these tasks, as they are by many other multinational companies in a great number of productions, consumptions and exchanges, so that domination changes its principles, and groups, which have not the slightest democratic legitimacy, assume the place and functions of States. The danger is that groups or multinational companies take the place of Theuth and mock the old lessons of the Thamuses that do not have powers to match their discourses.

6. We began, like Socrates, from a principle of hostility between technique and politics, rather than from a pre-established harmony. We have tried to show that there is no reason to imagine them as adversaries. Democracy must not behave as an obstacle to technical development, even if its function is to question it, or ask for explanations and, if necessary, try and rectify some behaviours. The idea that some doctrinarians of the seventeenth and eighteenth centuries had of democracy can only constitute an ideological obstacle to it, but it does not express the reality of democratic states in the twentieth and twenty-first centuries. Our meditation on democracy has concluded in an antinomy: on the one hand, a thesis, which is well past its sell-by-date and moribund, describes a State in which each citizen would willingly unite, and the laws of which would be rooted in a general will conceived as a contract; on the other hand, the more realistic antithesis conceives of the State as an association in which every political unit folds back upon itself. Far from having any rooting and foundation, laws are contiguous constructs organized in patterns which contain crossings often more Zenonian than axiomatically derived from some substantial will. Democracies have nothing to fear from the development of techniques, particularly from artificial intelligence, even though the one and the other are not spontaneously harmonized.

Moreover, as early as the infancy of artificial intelligence in the seventeenth century, and still more nowadays, many techniques were in competition, from the intelligence which seems to form one body with human thought (following a Pascalian or Bayesian model) up to the production of complete listings of situations which are inaccessible to human thought (that may be called Fermatian, to go back in time no further than the seventeenth century).[40] We think, with G. Simondon, that

[40]Nicholas Carr shows this [9] when he distinguishes at least two types of strategies in the programming of computers: that which aims at exhausting all the possible solutions of a given situation, and that which rather follows the choices of human intelligence and seems to support it. Carr cleverly shows that it is not necessarily the second type that is the most efficient. Deep Blue, the computer that defeated Kasparov in 1997, followed the first strategy. In a way, this is an old story that has existed in mathematical strategies themselves. The history of probability calculus shows it, as Fermat's complete enumerations, the more selective trees of Pascal, which calculates

'philosophical thought must really accomplish the synthesis' between the technical world and what he called *religious thought*,[41] but what we prefer to call *political thought*.

However, what is certain is that, even if the technical networks compelled us to project ourselves towards the past and the future, they place us before a double dead-end. The *first* is the impossibility of giving sense to the dream of coming back to some supposed Adamite humanity, which would be devoid of machines and of the technical world – such an origin may be imagined, but it cannot exist: we can never know what the feelings and thoughts of men of the past were. The *second* is the possibility of projecting the future as if we could know it. That condemns us to the illusions of post-humanist phantasms. Not that man cannot be coupled to machines and so lose his identity, but none of his lost identities would be a natural identity. Our identity has always been mobile and will always be mobile, while its past – though it may seem necessary – has always been contingent.

It should also be stated that the difficulty we meet in the definition of what is called *democracy* nowadays, and of which the new technological deal, in particular, has erased many frontiers, gives no reason to blur all distinctions between the States which try to be democracies and the States which make no such attempt. There are countries where opponents may express themselves without risking their freedom and their life, and countries where this is impossible. There are countries which are so unequal that there is nearly no chance, when one is born in a 'bad' social class, to share the lot, condition and living standards of those who effortlessly take advantage of the social benefits of the 'better' classes. Undoubtedly, we must avoid an easy and not very convincing idealism, but it is necessary to avoid relativism, which is no less easy and no less dangerous.

6.8 A Few Conclusions

The first conclusion takes the form of underlining a particular difficulty of the ethics of intimacy: it emphasizes '*be*' too much, and '*must-be*' not enough. Admittedly, the stupidity of moralism is always due to a too feeble and too peremptory an ontology

the expectations of players, are opposed. A century later, Bayes, in order to solve his famous game problem, played on the reflection of who speculates rather than on an objective enumeration of all the possible solutions of a given situation. It is by following a Fermatian scheme that, for example, the theorem of the four colours could be formulated: 'en explorant de façon systématique toute une multitude de combinaisons qu'il eût été fastidieux d'énumérer à la main' ([14], p. 58).

[41] [46], p. 238: 'La pensée philosophique doit accomplir réellement la synthèse et elle doit construire la culture, coextensive à l'aboutissement de toute la pensée technique et de toute la pensée religieuse. (...) La culture doit réunir réellement toute la pensée technique et toute la pensée religieuse. (...) La pensée philosophique saisit et traduit la portée de cet intervalle; elle le considère comme positivement significatif, non comme un domaine statiquement libre, mais comme la direction définie par la divergence de deux modes de pensée; (...) la pensée philosophique prend naissance au long du devenir divergent pour le faire reconverger'.

which, instead of looking carefully, already knows that the machine only knows signals, but not signs as we know them. One makes oneself ridiculous when one champions a pure symbolism, as if things themselves were not more mixed, and as if their existence bore no relation at all to these abstract distinctions. There is nonetheless a point at which the ethics of intimacy as we have tried to defend it becomes problematic in relation to the classical morality inherited from the Enlightenment. It fills up being, so that it does not leave the void which, for some centuries, from Rousseau to Sartre, Kant and Fichte, was allowed to man, and within which he could define himself. Man is not what he is, he must become it. The concept of *intimacy's* shortcoming is that it essentializes the individual. This is the only point on which the concept of the *person* seems to retain a decisive advantage over that of *intimacy*, which owes too much to being and not enough to the nothingness which deepens in front of that being or separates it from itself. Curiously enough, the robot and the linkage with the robot play the role of filling up the must-be to the point of making it being itself. For, if, as we have noted many times, the perfection of our acts, of our measures for example, is a phantasm produced by our acts themselves – by that of measuring, for example, which, by being performed, produced the norm according to which it measured – the acts produced by machines could attain a perfection that may no longer be only something to be fantasized. Maybe there is no mankind without that want, that lack. Does the machine not deprive the being with which it is coupled of the want that is necessary to existence? It can saturate with being, like a final written text which is not to be touched or edited anymore. Can we conceive of an ethics that would be thus saturated? The solution may lie in a deepening of what we have called *writing* or *engraving*, the temporality of which displaces rather than annihilates the difference between being and must-be.

There exists a more ethical answer, however: the concept of *must-be* may have a moral value, even though not a necessary one, and morality without duty may have been recommended, for instance, by G.E. Moore in his *Ethics* ([33]) ([34], pp. 53–4). If the moralities which include duties defend themselves well, however, the idea of duty is not an ethical concept, since for a good to become a duty that good must be considered as presenting an exclusive claim to obedience, or, if not an exclusive claim, at least a claim superior to all others. Ethics however, does not recognize any good superior to all others, except perhaps that of seeking to reduce conflict, which is proper to any diplomacy. It is not impossible that some play, some gap between being and must-be is necessary, but this is not exactly an ethical problem, except where all the actors in a situation considered from the ethical perspective pose the problem in terms of the 'must-be'. Ethics is not constrained, however, to adopt any such position, and, in our search for ethical categories, we must not be afraid to move away from moral categories which, if they were to be imposed, would radically undermine and destroy ethics.

Secondly, one can read in the Bible a beautiful text on numerical proverbs: 'Three things are too wonderful for me, four I do not understand: the way of an eagle in the sky, the way of a snake on a rock, the way of a ship on the high seas, and the way of a man with a girl.'[42] In the text, no comparison is made between the four traces quoted, and as its author seems to have some difficulty in distinguishing between three and four, I will add another trace, which is enigmatic for me, and which could be a fourth or fifth trace which amazes me as much as it amazes many others: that of men coupled with robots, which is the case in relation to a high number of disabled, very old or mentally-ill people. What will the sign of the future, of our future, be? What will robots, or the couples which give them a place in constituting a new individual – provided, at least, that they still write traces in the sense in which we have understood it – write? I do not know, nor do I know what will be written in the void left when I am no more, when I am no longer able to write it – no longer there to allow a writing to go through me – or able to displace a sign or to read it. How could I know what will be engraved, if I do not even know how things are being engraved under my eyes, at the very moment I am living them? We should fear neither this learned ignorance nor this non-science, provided that they, and we, remain vigilant, and so long as there is a subject of being and becoming, and a subject of being which retains a sense of itself which extends itself beyond the present, without a frozen rigidity which would be impossible to bear.

There is no need to complain, as my friend J.Y. Goffi does, that 'le propre des critiques formulées en ces termes (en termes humanistes contre le transhumanisme), c'est qu'elles ne discutent pas les thèses transhumanistes dans leurs propres termes', adding the reproach 'qu'elles se placent d'emblée en surplomb et traitent les affirmations comme des symptômes'. ([15], p. 11) Could we not simply answer that transhumanism does not manage to express itself in the terms it recommends, and that it reproaches its adversaries for not adopting? Has there ever been a prediction that was realized as it was announced? The spokespersons of that movement only feign to be writing about the future, for they write, like everybody else, only in the present, or, if one prefers, they write about a purely imaginary future in the present. How could it be different? How could any statement straining towards an as yet unrealized future be anything but prone to wandering from what will be? Those who write transhumanist texts are in roughly the same situation as those who ask the dead to speak of themselves, or who ask men of the present or be men of the past or of the future. The arguments which Hume used to explain why we could not sincerely be interested in life after death apply here. Men are not usually interested in them. Transhumanists are interested in them, but they cannot, with their own words, properly answer the questions in which they are interested. The beyond they wish to speak of is as radical as death but would not be death, although, one can say, like Hume, that 'some new species of logic is requisite for [the] purpose [of understanding it]; and some new faculties of the mind, that they may enable us to comprehend that logic'. ([18], p. 406) Their beyond does not allow itself to be spoken any more than the beyond of death. The antithetic is prolonged in a sort of meta-dialectic which is no more soluble than the ordinary dialectic constituted by

[42]*La Bible*, [6], *Les Proverbes*, XXX, 18–19, Tome II, p. 1439. [7], p. 599.

death, otherness, the reverse of phenomena, or the possibility of projecting ourselves in time, none of which prevent either scientific work or practice.

This is why I would like to end with a paradox. I am wondering whether things themselves may not force us to do what we have refused to do in relation to intelligence and the diversions of the symbolic, and whether this is not a point on which transhumanists will turn out to have been correct: not through a direct discourse on things of which they know no more than others, but by precipitating, if they remain consistent – and they do not always do so – the end of an ethics that no longer allows us to think the present, and by accelerating, maybe against themselves, the desire for an ethics more in conformity with new times. Indeed, everyone knows how outdated and unsuitable Beauchamp and Childress's system has become, if it ever was suitable. Nonetheless, one often thinks and acts as if, by means of a few amendments and a few fictional constructions, one could yet be satisfied with these old categories. We could have taken utilitarian calculations more seriously. We may find ourselves in the situation the transhumanists push us recognize – and this may be the most important thing that they have to say – that the degree of technical development forces us to abandon a 'personalistic' system of thought which has never functioned without inconsistencies, but which created the illusion that it was possible to adopt and implement it. It is quite disagreeable to *be pushed* by things themselves to do what we have not *wanted* to do. Those who are moved by, and in any case want to move us, with visions of unthinkable futures – as if futures, whatever they may be, had ever been anything else than unthinkable, at any time whatsoever – may serve as a diagnostic symptom that it is time we changed our ethical categories, or that it is necessary, in any case, really to work on their review. I have tried to establish that intimacy could be one of the possible categories in that modification, and that though it may not be that new, it does not in fact owe very much to the traditional categories to which it seems closely related.

Thus, and though this may still seem paradoxical, robotization is not an obstacle to intimacy, which is likely to resist the revision or even the loss or repudiation of its sister, neighbour and rival concepts. What makes us think it is a paradox is the bias which assumes that intimacy is quite a humanist value, and one which consists in contrasting the human person endowed with purpose or moral destiny with the machine, and therefore in contrasting ethics with technique. This conception is wrong, since ethics, when it is correctly conceived and not intuitively understood in terms of a particular moral mode, poses the same problems as technique. It continuously requires the articulation of principles, and the experience of that articulation never works without frictions and tensions in the circumstances in which it acts, and it is always to be done again. This is how the technical object only progressively becomes, in a way, indefinitely concrete. This is also how any ethical 'solution' must always be thought anew in every new circumstance. One will not always render mankind the great service one thinks one is rendering it by endeavouring to support it against techniques. On this point, G. Simondon's beautiful lesson still applies.

Bibliography

1. Bentham, J. (1827). *Rationale of judicial evidence* (5 Vols) (Vol. I). London: Hunt & Clarke.
2. Bentham, J. (1843). Constitutional code. In J. Bowring (Ed.), *Works of Jeremy Bentham* (11 Vols., Tait, Edinburgh, 1838–1843, Vol. IX. 1995). Reprint of the 1843 edition by Thoemmes Press, Bristol.
3. Bentham, J. (1843). Plan of parliamentary reform. In J. Bowring (Ed.), *Works of Jeremy Bentham* (11 Vols., Tait, Edinburgh, 1838–1843, Vol. III).
4. Bentham, J. (1970). *Of Laws in general*. London: Athlone Press.
5. Bentham, J. (2010). *Of the limits of the penal branch of jurisprudence*. Oxford: Clarendon Press.
6. Bible. (1959). La Bible, L'Ancien Testament, NRF Bibliothèque de la Pléiade, Gallimard, *Les Proverbes*, Tome II.
7. 1989. *The Holy Bible containing The Old and New Testaments*. Cambridge University Press.
8. Boyle, R. (1996). *A free enquiry into the vulgarly received notion of nature* (1686). Cambridge University Press.
9. Carr, N. (2019). Un grand maître humilié redresse la tête. In n° 94 of *Books*, in February 2019.
10. Cléro, J.-P. (2019). *Rethinking medical ethics*. Stuttgart: IBIDEM.
11. Descartes, R. (1991). *Principles of philosophy*, IV, art. 203 (V. R. Miller & R. P. Miller, Trans.). Dordrecht/Boston/London: University of Western Ontario, Kluwer Academic Publishers.
12. Descartes, R. (1996). Le Monde. In *Œuvres de Descartes* (Vol. XI). Paris: Vrin.
13. Foucault, M. (2015). *Les Mots et les choses*. Paris: Gallimard, Bibliothèque de la Pléiade.
14. Ganascia, J.-G. (2017). *Intelligence artificielle: vers une domination programmée?* Paris: Le Cavalier Bleu.
15. Goffi, J.-Y. Transhumanisme. Academic version in Kristanek M., *L'Encyclopédie Philosophique*. http://encyclo-philo.fr/transhumanisme-a/
16. Harari, Y. N. (2014). *Homo Deus*. Cambridge University Press.
17. Hottois G., 1979, L'inflation du langage dans la philosophie contemporaine: Causes, formes et limites, Brussels Université libre de Bruxelles, Faculté des Lettres.
18. Hume, D. (1992). *Of the Immortality of the Soul*. In D. Hume (Ed.), *Essays moral, political and literary*, 2 volumes, ed. T. H. Green & T. H. Grose, Scientia Verlag Aalen, vol. II. 2001, Essais moraux, politiques et littéraires (pp. 683–692). Paris: PUF.
19. Hume, D. (2011). *A treatise of human nature* (D. F. Norton & M. J. Norton, Ed.). Oxford: Clarendon Press, 2 Vols.
20. Jankélévitch, V. (1983). *L'irréversible et la nostalgie*. Paris: Flammarion. Particularly, Chap. V.
21. Kant, I. (1996). *Practical philosophy* (M. J. Gregor, Trans., & A. Wood, Intr.). Cambridge/New York/Melbourne: Cambridge University.
22. Kant, I. (1997). *Critique de la raison pure*. Paris: Quadrige/PUF. Kant, I. (2000) *Critique of pure reason*. Cambridge/New York/Melbourne: Cambridge University Press.
23. Lacan, J. (1978). *Le Séminaire, L. II, Le moi dans la théorie de Freud et dans la technique de la psychanalyse*. Paris: éd. du Seuil.
24. Lacan, J. (2004). *Le Séminaire, L. X, L'angoisse*. Paris: éd. du Seuil.
25. Leibniz, G.-W. (2001). *Méditations sur la connaissance, la vérité et les idées*. In *Opuscules philosophiques choisis* (trad. par P. Schrecker). Paris: Vrin.
26. Mauss, M. (2012). *Essai sur le don, Forme et raison de l'échange dans les sociétés archaïques*. Paris: Quadrige PUF.
27. *Medical robotics*. edited by J. Troccaz, Cachan, Lavoisier, 2012.
28. Merleau-Ponty, J. (1945). *La Phénoménologie de la Perception*. Paris: NRF-Gallimard.
29. Merleau-Ponty, J. (1967). *La structure du comportement*. Paris: PUF.
30. de Montaigne, M. (1927). *The essays of Montaigne* (J. M. Robertson, Trans.), 2 vols. London: Oxford University Press, Humphrey Milford.
31. Montesquieu, C. Secondat, baron de. (1951). *L'Esprit des lois*, L. II, Chap. II. In *Oeuvres Complètes*, II, pp. 239–244. Paris: Gallimard.

32. Montesquieu, C. (1989). *The spirit of the laws* (A. M. Cohler, B. C. Miller, & H. S. Stone, Trans & Ed.). Cambridge/New York/Port Chester/Melbourne/Sydney: Cambridge University Press. On the Net, one reads the book in *The Complete Works of M. de Montesquieu*, London, T. Evans, 1777, 4 vol., vol. 1, all.libertyfund.org/totles/montesquieu-complete-works-vol-1-the-spirit-of-laws.
33. Moore, G.-E. (1912). *Ethics*. London/New York: William and Norgate/H. Holt and Cy.
34. Moore, G.-E. (2019). *Éthique*. Paris: Hermann.
35. Pascal, B. (1995). *Pensées*. London: Penguin Books.
36. Pascal, B. (2004). *Les Provinciales. Pensées*. La pochothèque, Le Livre de Poche/Classiques Garnier, Paris.
37. Philonenko, A. (1997). *L'oeuvre de Kant – La philosophie critique*, tome 2, *Morale et Politique*, 3rd ed. Paris: Vrin
38. Plato. (1984). *Epinomis*. In *Plato in twelve volumes*, XII. Cambridge, MA/London; Harvard University Press/Heinemann W.
39. Plato. (1990). *Phaedrus*. In Plato, *Euthypro.Apology.Crito.Phaedo.Phaedrus* (H. N. Fowler, Trails.). Cambridge, MA/London: Harvard University Press.
40. Rousseau, J.-J. (1964). *Du Contract Social ou Principes du droit politique*. In *Oeuvres complètes*, Tome III, pp. 347–470. Paris: NRF Gallimard, Bib. de la Pléiade.
41. Rousseau, J.-J.. *The social contract* (J. Bennett, Trans.). earlymoderntexts.com/assets/pdfs/rousseau1762.pdf
42. Rousseau, J.-J. (1969). *Émile ou de l'éducation*. In *Oeuvres complètes*, Tome IV. Paris: NRF Gallimard, Bib. de la Pléiade.
43. Rousseau, J.-J. *Emile or education* (B. Foxley, Trans.). J.M Dent & Sons Ltd/E.P. Dutton & Co: London & Toronto/New York.
44. Shapin, S. (2019). '« Homo deus » really?' In *Books*, Feb. 2019, n° 94, pp. 30–35. S. Shapin's article, translated by D. Veaudor had already been published in the *London Review of Books* on 13th July 2017.
45. Sicard, D. (1999). *Hippocrate et le Scanner*. Desclée de Brouwer.
46. Simondon, G. (1958). *Du mode d'existence des objets techniques*. Paris: Aubier.
47. Spinoza, B. *The ethics (ethica ordain geometric demonstrata)*. Trans. from the Latin by R.H.M. Elwes, globalgreyebooks.com.

Chapter 7
Conclusion: Is Meeting a Reality or a Fiction? – A Few Epistemological Reflections on the Possibility of Constituting a Concept of *Meeting*

Abstract In the wake of the previous chapter, and taking the opposite of complaints and unquietness -though without forgetting them- of a world that fears having to deal only with robots and never more with beings of flesh and blood, we suggest that linkage and promiscuity with machines do not « destroy mankind », but rather take part in its irretrievable and irreducible wanderings. Complaints, that intend to be more or less philosophical and which are heard here and there, often seem to us the effect of mere intuition, without leading the least authentic research about the real damages caused to « man » -if the notion remains indeed valid- nor providing the least solution.

We notice that one of the passions unceasingly awaken by the linkage with machines might be *astonishment*, *wonder* resulting from the event of meeting. It is a misconception that the notion of *meeting* must be reserved to relations between humans only and valued provided it be in a unique intersubjective context. The machines -those that save, support and facilitate life- may be « met ». Descartes made of the *astonishment* the first of the passions of soul; the reasons to make this choice may still be ours even beyond what that great philosopher might think.

A new approach to the notion of *intimacy* and to the notion of *meeting* as two main categories of ethics seems necessary to be activated.

Keywords Analytics and dialectics of the notion of meeting · Fiction · Intimacy · Meeting · Passion · Robot

If, after the deeper meaning that the concept of *intimacy* has received from its confrontation with a context where the insertion into a world of machines – and of very particular machines that, together with our contemporaries, we readily call 'intelligent' – we want to conclude by investigating the concept of *meeting*, it is because of a situation which could not escape any researcher who takes an interest in medical ethics – that of the new relations that the linkage of each and all with machines and, maybe more specifically even, that of the help that computers and robots are supposed to bring to the elderly in our community.

J.-P. Cléro, *Reflections on Medical Ethics*, Philosophy and Medicine 138,
https://doi.org/10.1007/978-3-030-65233-3_7

Though the authors who tackle this issue do not all fall into catastrophism, and though they 'imaginent aisément tout le bénéfice que le consommateur tirera de ces technologies',[1] they almost all end up warning against the risks entailed for our privacy, not only because of 'vulnérabilités induites par le réseau, lorsque d'habiles programmeurs mal intentionnés franchiront les barrières de protection et feront irruption chez vous dans vos objets familiers, par exemple dans votre voiture, dans votre montre, dans votre réfrigérateur, (...) voire, si vous en avez un, dans votre stimulateur cardiaque,' ([9], p. 73) but also, and more fundamentally, in relation to that perception of danger that still remains engraved in a traditional morals, for the reason that Serge Tisseron indicates, and which we would like to explore to finish this book: 'Les robots qui arrivent sur le marché pourront non seulement nous aider dans notre vie quotidienne, mais aussi nous faire la conversation en comprenant nos émotions et nos intentions, et nous répondre avec des intonations et des mimiques adaptées. Et, très vite, à force de les fréquenter, nous risquons de penser qu'ils sont des compagnons bien plus agréables et faciles à vivre que les humains.' [20]. J.-G Ganascia, who quotes from this text, adds a comment: 'The diffusion of pet robots is faced (in France in particular) with cultural obstacles that are not to be found in other countries, especially Japan.[2] The fact that the animal is a virtual one and, above all, the apparently regressive character of whoever exchanges with what is considered to be a 'toy', lead to these therapeutic uses of pet robots being discredited. A psychoanalyst like Serge Tisseron even mentions in *Le jour où mon robot m'aimera* ('On the day that my robot will love me') an empathy he calls artificial because it leads to identifying oneself to robots. He sees at the same time an oxymoron and a major risk of dehumanisation.' [9] This assessment is shared by our contemporaries who fear that robots, especially when they have a humanoid form, can reduce the capacity of elderly people to feel at ease with and to develop attachment to real human beings.

One must nonetheless ask the question: Why should the meeting be more of a lie, more of a fallacy, when it is with robots rather than with 'real' men? Why would that encounter be less ethical, and why would it oblige less those who perform it? The answer is usually quite readily given: but it is misleading, and we should beware being misled. Thus, to the question – Why would it be bad to use robots to keep old people company? – the usual answer is: because one then makes believe in a meeting which does not actually happen and which just puts a person in the presence of an image of himself that he does not recognize as such, and that he simply mistakes for someone else or for the image of someone else. As if one could not be misled by

[1]Those systems ensure a presence. As they move, they can follow the person and answer his questions, when they are asked aloud. They also allow the family and friends to keep in touch remotely. This is in any case one of the arguments used by the developers of pet robotics. Let us recall that patients who suffer from dementia repeatedly ask the same question of their increasingly exasperated relatives, ten, fifty, or a hundred times. A pet robot with a conversational agent answers the repeated questions with endless patience, relieving the care givers and relatives from avoidable stress and reducing the demands on their time.

[2]Authors often note the great success in Japan of these pet robots – which take the form of little seals or dogs – with the elderly.

one's peers! Or, as if being deceived by one's peers, by other human beings, were always more worthwhile, for a human being, than being deceived by other beings or being deceived by themselves, or by the mirror with which those other beings present them!

Many think that there is only a meeting when the action one is performing towards a being triggers a 'similar' reaction, or an approximately symmetrical one, from that being. Only a fairly close resemblance with the partner of the meeting would guarantee that the meeting is ethical. But why should one grant a privilege to the meeting with such an other? Why would it be more ethical to meet the other – whom we will call 'human' for want of a better word – than to meet a thing, even one endowed with complex capacities? We mention complexity because the wealth of meetings with things other than human beings, in particular when we are coupled with them, results in their not exclusively happening with beings that look like us, and which send us back an image that is quite similar to the one we send them. We do not see why there should be less illusion in the presence of human beings who 'resemble' one – with all the ambivalences that are attached to that 'resemblance': do other human beings all look like us? From what perspective can it be decided whether that resemblance is present and what content it should be given? Does one not meet a certain number of people in whom one is reflected, but nothing more?[3] Why would such a meeting be an ethical failure? Why would one illusion be worth more than the other? Who can decide it, and based on what criteria?

An element that C. Salmon notes in Kafka ([18], p. 19) puts us on the right track. Refusing to go on writing letters to Milena Jesenská, Kafka explains that 'les baisers écrits ne parviennent pas à destination, les fantômes les boivent en route'. Thus he draws a distinction between two types of techniques, those which really lead men away from one another and bring them together, like the train or the plane, and those which only create phantasmatic relations among men, like the wireless telegraph or the telephone. Kafka's classification is far from being satisfactory, but it is indicative enough, and we have no doubt that he would have classified – if he had known them – our computing relations under the heading of phantasmatic relations, which lack a relation with real otherness, that of one-to-one, in the flesh, if we may say so. The remark is undoubtedly suggestive, but it does not make it possible to avoid an analysis of meeting, and it encourages us to take up the problem that Descartes did not hesitate to pose at the beginning of the treatise *The Passions of the Soul*, which is a book of physics as much as a book of ethics, since the passions are its material, and admiration – at the heart of which lies the meeting with some surprising thing[4] – is the first of them.

---

[3] Has Lacan not amply shown, in his analysis of *Symposium*, that 'l'être de l'autre dans le désir n'est point un sujet. L'*éroménos* est *érômenon*, au neutre, et aussi bien τα παιδικά, au neutre pluriel – les choses de l'enfant aimé, peut-on traduire. L'autre, en tant qu'il est visé dans le désir, est visé comme objet aimé' ([12], p. 66).

[4] [7], p. 373. <«Wonder. When our first encounter with some object surprises us and we find it novel, or very different from what we formerly knew or from what we supposed it ought to be, this causes us to wonder and to be astonished at it. Since all tis may happen before we know whether or

The problem is important as it makes us consider in a new light the ethical discredit of the linkage of a 'human' being with machines. Whom is one meeting? What is one meeting? Is it only possible to meet other human beings? What makes it a meeting? What prejudice would make one admit that only meetings with 'human' beings have an ethical scope or weight? When I am using an object or a thing – whether I know it or not – as a mirror in which I am reflected, why is there no meeting? Why would that type of narcissism not be a meeting and why should one refuse to give it any ethical value?

7.1 *Meeting* Is, at First Sight, a Very Vague Word

Is there a concept of *meeting*, or, in speaking of *meeting* is one using a word that gathers a disparate number of situations, dispositions and values between which it is not at first easy to see the rational link? Is the *meeting* something which would *allow us to discover*, as if the signified of the word indicated a unity of essence which should be grasped, or does the word always mark a *construction*, whether it is dissimulated, admitted or stated as such? Of this last alternative – discovery or construction – the former idea can represent the intuition of the essence of meeting as the backbone of the construction we are talking about, while the latter can do without that intuition, or even destabilize it, by asserting that anyone who thinks he is paying attention or pretends he is paying attention to the concept through the word would only be hiding a fabrication, by making a clandestine sculpture that he does not admit publicly or even to himself. Should one listen to the word, or should one build it by enclosing within it as many as possible, or even all, of the senses that one would have detected before, or at least some of them, in the same structure? *Meeting* is first a signifier, which seems to embrace a great number of signified which are not well articulated, or, to put it more prudently, of which one does not know whether they are well articulated. One has only to think about the different uses of the word by reading *Le trésor de la langue française* (The treasures of the French language) published by Klincksieck, to become aware, admittedly not of the exhaustive totality of those meanings, but of many of them. Should one, to create the unity of the word, accept the pruning from it of a certain number of meanings? But on what foundation and in what capacity should one do this? Or should one try to embrace, as much as is possible, all its uses, without exception, at the risk of making the word *meeting* an operator of meaning that is almost as broad and indeterminate as the word *do*, and which can go with almost any verb and express almost any activity?

If we wanted to do an analytics of *meeting*, that is, an analysis of each of the elements of which it is made up, in the most exhaustive way, we should start with

---

not the object is beneficial to us, I regard wonder as the first of all passions. It has no opposite, for, if the object before us has no characteristics that surprise us, we are not moved by it at all and we consider it without passion» ([6], p. 350)>.

elucidating the issue of what the object of that analytics is. Is it the 'very thing', the meeting itself, provided it exists? Is it the intertwining of the uses of the word? Should we trust the word, which, thanks to the unity of its signifier, seems to be announcing a unity of the signified, the reason for which initially elude one? Should we let ourselves be guided by the presumed unity of the word in which we should believe? Or should we cut into the word, state its principles, without hiding them behind a supposed listening, the experienced passivity of which hides a real selective activity?

Undoubtedly, cutting into the meanings is to be forbidden because it is theoretically hypocritical, but we might agree that the so-called guidance given by the presumed unity of the word may have the reverse effect of that which is sought, and that it might threaten to inflate the concept of *meeting* or the signified 'meeting' to bursting point: one talks of lines or masses that meet as much in geometry or physics, as in law, ethics or politics one talks of persons who meet in particular or private relations or of heads of state who meet.

An ANALYTICS of the meeting may try to build the unity of the different meanings without exception, in particular without being content with the problem presented by the meeting with the other, which we tend, as we have seen, to favour spontaneously in our analyses. A cut that would lead us to confine the meeting to intersubjective relations would condemn us to overlook it at once. It would from the beginning mar the concept that is being analyzed. When we say that one line meets another in geometry, or that one body meets another in physics, we do not have any reason to think that this use of *meeting* is more metaphorical or figurative than that which is considered in the intersubjective sphere and which is often considered, from the outset, as the more real or primary sense of the word. The choice of what is held to be real or primary must be considered to be suspicious if it does not give its reasons and thus performs an arbitrary selection.

At the end of that construction, be it inspired by an intuition or remain blind or symbolic,[5] it is clear that one will have produced a certain number of contradictions, admittedly linked to the diversity of the meanings of the concept of *meeting* or *encounter*, but also to the diversity of the qualifications one can impose in relation to it. It is not impossible to number those contradictions. At least the project of undertaking such an enumeration is not meaningless. The display of the contradictions constitutes what could be called an ANTITHESIS of the meeting or of the concept of *meeting*.

Any such an antithesis may be disastrous if the concepts which are used to establish the contradictions let the essentials of what appeared to be an idea slip away. If one believes that that idea is an illusory lure, the antithesis will be taken seriously enough to be granted a reality coefficient that is not granted to the idea. If one believes, on the contrary, that the idea is not an illusory lure, then the antithesis, with its set of contradictory ideas, will appear as a game of fictions: the tearing apart of the idea then, seems to come from the conviction that one does not know how to

[5]As Leibniz would have said.

conceptualize its unity. How is one to conceive of the unitary and particular inscription of the word *meeting*?

One will note – and this will form a common thread for us – that the word *rencontre*, which appeared quite late in French, is difficult to translate into other languages. It has no strict equivalent in Latin, even though each of its elements does. Its Latin translations give the retrospective impression of dismembering the idea of *rencontre* by rendering it, according to need, as *occursus*, *adversus*, *concursio* or *inventio*.[6] The word constitutes a particular alloy characteristic of the French language, which manages to conciliate, in the same word, both a surprisingly broad spectrum of meanings and an absolutely singular flavour. The word has been charged, over time, with all sorts of meanings, which have made its signified drift in an incredible way. If one wants to translate it into English by *meeting* or *encounter*, or in German by *Begegnung*, *Begegnis*, *Zufall*, *Einfall*, or *Treffen*, or by one of several words derived from *entgegen*, one will always seem to retain only part of the word, and not its entirety. Bentham, in *Table of the Springs of Action* (1815) [2] [3], which lists the motives of action, noted that the list, which he compiled only in English with the addition of a few foreign words, would have been quite different had he had to compile it in another language.[7] The difference in the schemes of actions and in the ways of feeling affect not only the modes according to which they can be conceived. It also affects the modes according to which one lives them inside one's walls, laws, institutions or modes of social organization. The *meeting* is one of those schemes, even though Bentham does not write about it. In French, it expresses not only something that is to be thought, but which is also to be lived as something that lacks a common measure with that which is to be thought and lived by words which are apparently closely related but which belong to foreign languages.

Our hypothesis therefore, is that comparing languages, which is very effective in translation, may provide an efficient point of reference that would not be provided by logic, which allows the idea which other vernaculars or discourses attempt to express fall into its own contradictions. Curiously enough, the points where translation runs into difficulties can pose tough problems, which do not let themselves be easily penetrated or dissolved, but which permit reflections on meeting to be built which has a greater chance of approaching truth than that which is obtained by an analytics and by the attempt to resolve a dialectic.

Without ignoring the dangers of working in a fragmentary way, I would like, under each of the approaches I have enounced – analytics, dialectics (or antithesis),

---

[6]Greek gives the same impression since to translate the French word *rencontre*, one would have to turn, depending on the contexts, to words that are as diverse as ἀπάντησις, ἔντευξις, ἀπ-αντάω (going to meet), σύνοδος, σύγκρουσις (shock of two objects), συμbολή (shock of two armies), καιρός (occasion), τύλή (meeting) derived εὐτυχία, δυστυχία.

[7]I have translated *Table of the Springs of Action* [2] into French – *La table des ressorts de l'action* [3] – but, as far as I have been able to ascertain with the help of Michael Quinn at the Bentham Project in London – and despite the expressed wish of Bentham himself, who did not, however, expect much from the French in this regard, it has not been translated into any other language.

and the taking into account of the *meeting* sign in its particular engraving – simply to note, under each head, a very few aspects of the question, which will suggest, in an overview-like conclusion, how the rest could have been dealt with.

7.2 Drafting an Analytics

Drafting an analytics is exploring the *meeting* as a set of categories, a way of making use of time and of deploying a space. One should not rush to say what the subject of the meeting is, as if it were a being that is substantially given, by favouring the persons, endowed with conscience and will, who encounter others. One would be quite wrong to think that a subject or that subjects who stand(s) before one are the only beings which could ever be met. One can meet things, an animal, animals,[8] landscapes, towns, situations which imply several actors none of whom especially stand out. Geometrical beings can be met, as can signs, abstract and fictious beings which have been created on purpose. The meeting that happens then is not more illegitimate, nor less important, than that between persons or 'human' beings, who seem, each to the other, to be obvious candidates as parties to a meeting. The mistake is to believe that the meeting is necessarily a symmetrical relation, and that one only ever meets beings which are capable of meeting one, especially those which experience the meeting in a mode that is very similar to one's own. But one can very well meet someone or something that does not have the slightest idea that one has met them: an actor on stage that I watch playing a part, a musician who plays for an audience among whom I am present, a politician who delivers a speech listened to by many people and who does not particularly care about me. Incidentally, one can less easily meet a musician than a piece of music or a musical work, less an actor than a play. It is very difficult to determine exactly what is being met in a given situation. Meeting does not necessarily imply a sort of delving into the otherness of the other, or into the thingness of things, or even into the otherness of the thing, as when one gives a thing some features that are human, or when it is reproached for not having those human features. The being that I meet can play a role that he does not intend to play in the scenario I involve him in, into which I drag him unawares, just as I may be completely ignorant of the role that I am interpreted as playing on the stage he offers me.

[8]One can read, from that perspective, the beautiful beginning of a text that is quite remarkable for our own text and which is entitled *Première Rencontre. Le Cheval et l'Homme*: (*Vingt Écrivains Rêvent*), [1]. Its author makes one feel how the horses introduced by the conquerors of America were perceived as strange beings by Indians, who could not identify them as flesh-and-bone beings. Real animals may be met as artifices, as some sort of toys in the reality of which one does not believe. In that meeting of Native Americans and Europeans, things and animals were at least as strange as human beings and were lived, in perceptions themselves, as beings unquestionably existing and yet unauthentic. Such discrepancy, which is linked to the collision between an expected experience and what is really seen but difficult to believe, is an experience of meeting.

What does one mean when one says that *a parabola meets a straight line* or some other curve, or that *a tangent meets a curve*? One would be quite wrong in thinking that *meeting* here is a word which is fuller of imagery but which carries the same import as *intersection* or *contact*. Admittedly, the meaning of these words is quite similar, but above all it indicates a difference in values. Just as it is at the intersection point of the parabola that something interesting happens, and not when the branches goes on into infinity, like straight lines, it is in the intersection points that one is mainly interested. Speaking, in that case, of a meeting is a valorization of the intersection points, since it presents them as points where something interesting happens, something which is being staged. Not everything is interesting in the life of a parabola. Only a small number of incidents are. The meeting, which dramatizes a cut, a proximity or some other incident, indicates less the very point(s), the very event(s), than the choice or the selection of the most remarkable points or most interesting positions. It is the same in physics. The processes are only considered interesting when an event happens. That a ball goes at a constant speed according to a straight line on a plane is without interest. The meeting of a ball with another creates an incident which is worth inspecting. It is at the moment when the inertia principle combines with other principles that that articulation is thought worthy of attention. Admittedly, the straight lines of the geometrician may transect one another without changing courses, but it seems that if one pays attention to things and phenomena, as is the case in physics, the meetings result in course changes. Let us be very clear: it is not the mathematician subject who meets a scenario – that could happen, but then that meeting should be analyzed differently – it is not the physicist subject who does the same with the subject of experience that we have just sketched. In both cases, it is a meeting that happens 'in front of' a subject, who is called as a witness and who only stages or facilitates it. Let us not forget that the blade of a knife, as a subject, may meet an obstacle in its cutting. Then it cannot function as a blade in the same conditions as before. The valorization of contingent characters is even more apparent in biology, where one speaks, for example, of the meeting of two gametes, in which the idea of luck or unpredictability, which dramatizes the process, intervenes.

Everybody will have understood that the component of subject in the concept of meeting is extremely complex and cannot be determined in advance. From meeting, to subject, from mode of meeting to mode of subject which experiments on it, the consequence is correct, not the reverse. It is the meeting that carves and sculpts subjects for itself, not the reverse; so much so that even though subjects may think they do, they never meet the totality of the subject of the other, but only some profile of that other, and they are not met themselves in their entirety, but also through profiles, in a game of which they do not master the entirety. The meeting cuts the subjects into tiny pieces like Harlequin's coat, or in indefinite profiles or modes, and it elects one among myriads of others. If, now, one looks on the object's side, one finds the same difficulty. There is no meeting without an experience of discrepancy between what one expected to see and what one sees in reality. Descartes does not meet the men he sees through his window, or rather of whom he only sees the hats and coats, as long as he does not think of doubting they are real men. As soon as the idea comes to his mind, however, that those hats and coats may be covering spectres or false men,[9] is one not rather concerned with, not so much a meeting, as the very conditions of a meeting? In other words, must there not be, for a meeting to happen, some failure of the usual conditions for the unproblematic operation of the working of an object, so that in a sense the conditions of the meeting precisely rush into the discrepancy that makes it possible? Could one not say, like Lacan when he talked about anxiety, that *the meeting is not without object*? But what is the *object* of a meeting? For example, what do people who meet want, and what is the object of an unwanted meeting? The object of the meeting is not the only empirical contact between two persons – flesh-and-bone ones, as some ethicists we have discussed sometimes say. Incidentally, a meeting can occur without any such contact. In a meeting, no one knows the object of the meeting, and there is no third being, either supposed or occupying a real place, which knows the object and holds the key to the meeting. Or, if one thinks that the other knows it, it is thus that one creates, for the meeting, a sort of solidity, a sort of fake density that can be misleading. The meeting seems to be its own sole object, an object that lets itself be experienced, and which is accepted as a transition. A meeting may have a witness that 'objectivizes' somewhat that happens, but it does not need such a witness to be a meeting, and that witness may easily be put into parentheses or disqualified in its objectivization or essentialization. The meeting becomes incidentally other when that 'objectivization' somehow becomes effective. One often feels that a meeting can only occur between beings that are capable of pursuing goals, but that is wrong, for the meeting, even if it is a finalized one, coincides neither with that finality nor with the meeting of two goals.

The important point here is how the meeting works. Its space is a space of the night rather than one clearly illuminated, if one accepts Minkowski's interesting distinction,[10] even though the meeting may happen in daylight. The space of the meeting works 'blindly'. It is that of trial and error, that is of some research that ends a little at random, since even though it brings goals into play, it does not know what it wants.

This remark encourages us to continue our sketch of an analytics in the sense of an aesthetics, in the Kantian meaning found in *Critique of Pure Reason*:[11] what are the time and space of the meeting? One often speaks of places of meeting, meeting points, meeting sites, as if those places, points or sites were to shelter the meeting, as

[9]We are referring here to the famous excerpt from the Second Meditation ([5], p. 25).

[10][13], p. 156ff. 'Lorsque l'obscurité fait place à la clarté, il ne s'agit point de ce que la lumière rend maintenant visible dans l'ambiance les objets qui s'y trouvaient antérieurement mais qui restaient inaccessibles à notre regard, mais de ce que le monde tout entier change d'aspect, obéit à des lois nouvelles, se peuple, entre autres particularités, non plus d'ombres fuyantes, mais de corps' (p. 157). Here Minkowski referred to a previous work, *Le Temps Vécu*, published in 1933 (reproduced by PUF with the same pages in 2013), ([14], p. 392), where the clear and black spaces are distinguished as modalities of the space and not as 'objective' sensory characters. His work ended on that image.

[11]That is, as knowledge of time and space, and not as knowledge of the beautiful.

if from the world outside, or as if the meeting were inherent to that place. In reality, things are, once again, more complex: The meeting may be linked to the search for, or even the attainment of, a coming closer together, a closeness. What is coming closer, or tends to come closer, in a space which is rather an affective space? If there were a representable meeting point, it is not *in* that space that the coming closer together occurs. The affective space only expresses itself in the space of representation. The meeting itself creates its own space. It gives properties to a space. It is in that 'making its space' that it happens. The space is then a meeting operator. It is essential to the meeting in the cafe, at a university, at the hospital, in Paris, etc. The place has an intermediary character. Here, a more detailed analysis would underline that the meeting space is a sort of discrepancy space: one thought one was going to see someone, according to some sort of inertia principle the reality of which was only phantasmatic, and things happened slightly differently. The meeting seems to escape two figures that seemed to define it. The first is the simple going through or coming into contact with the other – be it a human being, a thing or an animal – which one does not notice at all or which is not noticed by the beings that one thought one would meet. The second is due to the constitution of the other as a mirror-bearer, in whom one only recognizes oneself, or what is left of it. The meeting is something which happens between the going through or coming into contact with and the mirror: One's going through the other or being gone through by the other may, however, be meetings for that other that one neglects, or for oneself who is thus neglected.

As to the temporality of the meeting, it does seem that it is linked to the idea of *approach* up to the point of contact (or what one deems or has deemed to be such) or even to the more or less wide overcoming of the contact point that is being considered in the direction of the future, or of a past that it seems to have restructured. There is meeting when there is work; or more precisely, prospective work, the setting of a programme which is followed by the immediate effective beginning of work, even though it be modest,[12] without one knowing in exactly what direction that work is going, but being also such work as it is impossible for us not to start. Of course, here we understand work in the broadest meaning of the word, in the Freudian sense, which includes, but goes well beyond, the sense given to it in political economy.[13] Meeting someone that we have never seen before, whether we have contemplated meeting them or not, whether they have contemplated meeting us or not, is a meeting when *we imagine* that what we live in their presence

[12]This is why the Cartesian disposition which places *admiration*, the passion of meeting, before all the passions, as if they all owed it something, is so clever. In the meeting some process starts, even as a sketch or very modestly.

[13]That idea of *work* exists even when the meeting is between things. The meeting points of curves are indicators of work. The meeting between different elements is not a simple mixture of different elements: For a meeting to occur, the elements must take and take one another with a view to a problem and a solution or an answer. One could say, in an Aristotelian way, that the elements that meet are about to become the pieces of a *hypothesis*, which is the ontological moment of a proposition, and not only its simple character of statement.

is changing us or changing them. It is even probably real (at least it seems real to us) when we imagine that the meeting, when it is finished, has changed us, has changed our habits, that is, our way of thinking and living.[14] However, the *reality* of the meeting is difficult to assess. How can there be an impartial and efficient perspective to ascribe to a meeting its proper importance? Strong or enthusiastic feelings that it is of some importance may serve as a mask for operations that we do not want to know about: a change may be fantasized when we project onto the past some point of origin of a change in ourselves. The meeting may also be project in fantasy towards future when we believe that a more or less determined other that we are going to meet, and upon whom or which we think we may rely, can change us. Thus the strangeness of a meeting which is thought to be important as it is experienced may very well disappear after a few moments, hours or days. Conversely, the beginning of a meeting which will be efficient in making an important alteration in us may go unnoticed at the moment when the process is supposed to be happening.

It would be better to speak of beginning, in relation to meeting, or even better of origin, which does not imply a chronological beginning, which would be a dialectical and fallacious concept to use when talking about meeting. The origin does not imply a point or an instant rated as a beginning, since it spreads across many or even all parts of it, and is only believed to be a discrete starting point, which would not be ascertainable if one wanted to determine it.

The time and space of the meeting is that of a transmission of good or bad value, of value or anti-value, if one prefers. The meeting seems to be linked to the idea of seeing each other for the first time.[15] One would not say that one has met in the street someone with whom one usually lives, or someone whom one knows well. Or, if one does say so, it is because one did not expect to meet him there, or because one has discovered, on that occasion, an aspect of that person which had completely eluded one before. The fortuitous feature then slips from the person to the place where our paths met.

We are not going to continue this analysis of *rencontre* –in French, *encounter* in English – here, but should underline that it deploys a set of parameters concentrated in its name that is not found in the most common word *meeting*, so much so that the word has been retrieved in French, with, it is true, to counterbalance it, a deduction from the English meanings, or from those of the German *Begegnung*. The logic of the conceptual analysis is not exactly coincident with the logic of etymology. *Contre*, as in *encontre*, seems to be a substantive ancestor of *rencontre*, which is retained in English in *encounter*, a sort of opposition to something that presents itself

[14]Entitling his book, *L'Enfant à la Rencontre du Langage* [19], Dominique Taulelle indicates, from the very first sentences, that the acquisition of language demands tremendous efforts and a reflection on language from the child. One recognizes all the elements of the meeting, even though the author does not thematize the concept of *meeting* in his book. Such attitude is quite common: *meeting* can often be found in the titles of books, and seem thus to be summing up the experiences and providing a synthesis of them, but the author does not devote a single line of the book to explaining what he understands by *meeting*, as if the concept were well known.

[15]Though it is not impossible to talk of a first, second or third meeting.

as an obstacle. The word has incidentally a distant military meaning which it has retained, as the Klincksieck dictionary shows at the beginning of the eighteenth century. This *contre* is relatively illusory, in that the object is not immediately determined, but will be, since it is regarded as already on its way to be determined. There is in the word r*encontre*, *re-* of *renvoi*, of *renforcement*, of *regard*, of *recollection*, of *réflexion* (*reflection*), which means that the opposition of *contre* or *encontre* only happens in the reflexive repetition. I understand this word *réflexion* in the way Hume did,[16] as a *résonance* (resonance), without giving it any particular intellectual meaning. The aggressiveness of *encontre* is softened and, as it were, reversed under the form of some passivity or, at least, of an activity that does not know itself in the meeting.

We can already draw from this sketch of an analytics the idea that, if the meeting must play a decisive role in ethics – and especially in medical ethics – we cannot consider it without bias as exclusively a meeting with the 'human' other, and that instead the meeting may be that of thing-beings, simply because there is no thing in itself which would be apart from of the network that it constitutes in fusion, or, at least, in relation with the other 'human' beings, even though they have their own being of things apart from the other 'human' beings, no more than there are subjects in themselves and otherness in itself, comparable to ours though outside it. There is no reason to be afraid of a meeting of robots or of any other demonstrations of artificial intelligence. There is no reason either to consider – provided we acknowledge that meeting a robot can happen just as we can meet another human being or an animal or any other thing – that the ethical value of that meeting will necessarily be of an inferior quality to that with a human being or of an animal. The only true ethical problem is the one that we have already encountered (if the expression be allowed) in relation to *intimacy* and its trace, namely that of its relation to 'must-be'. We wondered then if the trace could, in itself, be a source of ethical value, and worried that a trace might very well not be morally qualified in the best of ways. We would like to ask the same question about *meeting*: are there not 'good' and 'bad' meetings? Successful or fruitful and disastrous meetings? In relation to meeting, we will answer in the same way as we did in relation to intimacy: there is no reason to confuse ethics and morals. Moral categories are not categories of ethics, and it seems to us that the categories of ethics are still to be discovered, and that they are still too dependent on moral categories even when they are not imagined as purely and simply coinciding with them. That morals is concerned with the good and the bad we will not dispute, but moral categories are not those of ethics. We owe to G.E. Moore the demonstration that the category of the *duty*, which implies that there is only one way of doing good, is not a category of ethics.[17] Is it not enough, for a category to be ethical, that is

---

[16]As when he talks of impressions of reflection (which are the equivalent of our passions or affects), unlike the impressions of sensation (which are the equivalent of our perceptions) [10].

[17][15], pp. 31–38. A French version of this text has been published by Hermann in 2019. One can read, pp. 34–5: 'It is true, in a sense, that whenever we act rightly, we are always doing our duty and doing what we ought. But what is not true is that, whenever a particular action is right, it is always our duty to do that particular action and no other. This is not true because, theoretically at least,

applies to a being or group of beings which, independently of contravention of laws, by its way of existing, does no harm, or does the least possible harm, to other beings' happiness and pleasure – be they individual humans or linked with machines (as they all are in one respect or another), singularly or in « brigues » <groups>, as Montesquieu would have said?[18] At least we should not be afraid of an ethical relation that did not happen in close connection with 'human' individuals, even though, of course, relations with exclusively non-human individuals would be problematic.

Moreover, the fear of seeing meeting extend to the thing or to the network that one constitutes with the thing(s) makes one commit the error which lies precisely in making a thing out of meeting, in failing to see that meeting is a mode of relation and does not lie in the thing itself. It is part of what Descartes calls admiration, which is not necessarily or exclusively admiration of persons.[19] One makes the meeting a thing more by disputing that one cannot meet things than by asserting that the meeting does not need to be only between human beings to be successful and to have an ethical meaning.

7.3 Dialectics of the Meeting

The previous analytical presentation throws us into countless contradictions which may be reduced to a game of oppositions along a few axes. Let us spell them out for the DIALECTICS of meeting that is the object of this research, and let us try to find their antitheses.

The main axes according to which the concept is divided and deployed, if one tries to make them abstract, are the following: (1) The meeting seems to be radically singular or inter individual, but it may also be collective, as in the case of a meeting between an individual with a group, or between a group and a group, directly (as between sports teams) or indirectly by delegation to some individuals who

cases may occur in which some other action would be quite equally right, and in such cases, we are obviously under no obligation whatever to do he one rather than the other: whichever we do, we shall be doing our duty and doing as we ought. And it would be rash to affirm that such cases never do practically occur. We all commonly hold that they do: that very often indeed we are under no positive obligation to do one action rather than some other; that it does not matter which we do. We must, then, be careful not to affirm that, because it is always our duty to act rightly, therefore any particular action, which is right, is always also one which it is our duty to do'.

[18]Unlike Rousseau who, in the chapter of *Contrat Social* [17] on the general will, teaches us to be wary of it, Montesquieu does not understand the word « brigue » <partial groups, in opposition with « global society »> in a pejorative way. See above note 36, p. 124–125.

[19]It is worth noting that the author who makes a big deal of *wonder* in his system of passions [7] [6] and, consequently, valorises the meeting, is the same author who, in *Le Monde* [8], opposes to the real and true world a world built that allows one to understand the first – be the two worlds out of step with each other. This idea of discrepancy is as much a principle of the Cartesian science of the world as it is the starting point of his system of passions.

represent them (as in meeting among the supporters of a same party, or parties, trade unions, interest groups, etc.). (2) The meeting seems to have to be reciprocal to become real: one meets someone who meets one. But we have seen that that is less than certain, and that one can perfectly well meet something or someone who will not care whether he has seen one, and who will not have been changed at all by the meeting even though it has changed one or would seem to have changed one deeply, and vice versa. (3) The meeting seems to defy all the codes which regulate the relations with the other, and seems to open a breach in those codes, but one may also consider it as radically coded, even though one did not previously see the codes that govern it. If it is often felt as an irruption, it is because one only sees it when there is a change of code or a breaking of the code. (4) In other words, the meeting seems to imply *immediacy*, but it is as difficult to refuse it *mediacy*: it has mediacy even in the folds of its French name, *rencontre*, which makes it a taking up of codes as much as a breaking of codes. (5) One could also talk about the game of *determination* and *indeterminacy*, along one of the axes: a meeting does not necessarily have a specific aim or object, especially if one understands aim or object to be empirically representable words, which does not prevent it from pursuing aims differently, for example affectively determined aims of which the actors are not necessarily aware. A meeting can have a content, however, which is quite determined at the beginning, at least in the heads of negotiators – which is the case in diplomatic or political meetings – and still remain a meeting, for it leaves some room for indeterminate elements, in the aims as much as in the means. When one says, for example, that two theses meet, one does not mean that they are identical, but one presumes that they have a few determinate points in common, though without any guarantee, and that is why one talks about a *meeting*. One often considers including elements of *hazard* – *de rencontre* is an equivalent in French of *de hasard* – as inherent to the meeting. It is supposed to be essential to the meeting not to expect to meet someone, and it is supposedly constitutive of it that chance be necessary to the existence of the meeting, or that it be related to its content, to what it offers. A meeting can nonetheless involve the exact contrary: it may be premeditated, composed, or built and present necessary frameworks: frailty and fortuitousness are then banished to its limits. (6) Lastly, one can conceive the meeting as a structural echo – let us leave undecided for now whether that structure is an individual or a collective one, all the more so as that division does not actually mean very much. In this case, the meeting does not change the structure much, at the utmost it makes it become conscious of itself, with all the ambivalences and illusions of the word *conscious*.[20] One can, conversely, insist on the deep cut or rift that the meeting introduces in the structure so that the whole must be rebalanced, 'taken anew', in a work that makes do with the other and restructures the same. The meeting often requires a 'taking up again', after a moment of dismay or hesitation. In the case where the meeting is an echo, that meeting is less real than it is declined in all the modes of fiction. On the contrary, if the cut is real, it is not the

[20]It may be better to talk, like Hume, about making the structure reflect on itself.

appearance of structural and deep processes: it is a condition of possibility of those processes, the continuity or identity of which may seem fictitious or illusory.

Of course it is out of the question to examine all those aspects one after the other here, given the space available, and the aim of this conclusion is not to do that work which will only be started here. I am not saying that, by working on each of those antinomies one could not define them differently, or find that some of them partially overlap, or consider the oppositions in a way that would allow for the two opposed words to be held together, by distinguishing levels of integration, for example, or even by making first one and then the other side right. I will only provide one consideration with a general scope here.

In all these cases (individual/collective, reciprocity/no reciprocity, code/breaking of the codes, immediacy/mediacy, structural phenomenon/breaking of the structure, what upsets habits/what can be repeated and 'reflected') there are folding phenomena. What is crucial lies at the fold of each antinomy, and it is the fold that must be accounted for in the game of what is being held to be real and what is being held as fictitious, so that these positions could be, in the cases we are dealing with, reversed. From each of the antinomies that we have reviewed, one can in turn hold the thesis as real while the antithesis stands in a fictitious relation with it, or hold the antithesis as real so that the thesis becomes the fiction. Thus, should one say that the structuring which allows the meeting is more 'real' than the accident that seems to be breaking it? In other words, should the 'real' structuring be taken as the starting point to account for the meeting? Or, conversely, should one start from its event, its accident, as being the only reality, from which one could project its structural past, and, beyond its birth, the other side of the structure, as being the constraints that have been obtained only by projection, as in topology? Would the meeting be the only mode of those transformations which affect the structures when an accident occurs, or is it then the very accident, the event, the interruption that makes it possible to give meaning to the differentiation of the structures which appear as fictions, as less real that the event itself?

One should not believe that starting with the event or the structures comes to the same thing, as if the reverse problems were perfectly symmetrical: they are not. Starting from the structures gives the meeting a status different altogether from that arrived at by starting from the event in its lived, experienced breaking. Will one ever establish the reasons for a convenient starting point? Of the event or the structure, which one is the most fundamental element, and how are we to solve the dilemma? The two possible options seem rather balanced, though they do not yield the same result. It is in that sense that one may speak of antinomy, or antinomies, since the six oppositions we have listed may be submitted to the question we have just asked.

7.4 A Method Inspired by the Benthamic Version of the Theory of Fictions

It might be said that if one poses the problem in this way, one is committing oneself to a method that seems to be very clear and finely determined, but that does not offer much hope of a solution. The particularity of the antinomic approach to problems is what is called *exhaustion* in mathematics, that is, that it consists in surrounding the object that one is interested in, by excess or default, in a game of opposed terms that are supposed to situate it so as eventually to measure it. The meeting is also at the heart of the oppositions between the individual and the collective, chance and necessity, the event and the structural, and the structural and the experienced. One may then, have the idea that, by that plural games, and determination through exhaustion, according to a by-default or by-excess framework, one may give an identity, a profile, to the object, provided the game is diversified enough. Putting the object of study, the *meeting*, at the centre of gravity of a whole game of contradictions that cross and recross it would thus be the best way of thinking about it.

In reality, since most of the contradictions which are distributed according to the axes that we have just reviewed are quite irresolvable – provided one sets aside pre-existing preferences linked to the more or less important force of one antinomic proposition compared to the other – one must confess that the critical and antinomic method clarifies, or even better makes some progress towards determining the issues, but does not make it possible to solve them. The method itself must then be assessed. It seems good, but it possesses the drawback of making the concepts it treats burst into a set of opposed concepts that contradict each other, but each of which is quite far removed from the core function of representing the concept under study. The falsity of such a dialectics lies in the fact that it thinks it starts from the concepts – as if they were independent from the words that are used in languages – and imagines that it corrects languages by use of those concepts, while the key to what should be done is rather to start from the very radical idiomatic singularity of languages, and to try as much as possible to act in accordance with that singularity. However, if one rejects the Kantian-style of analysis, is there another analysis, of a different style, which is not based upon a framework of opposed abstractions?

This is the point where the theory of fictions and its ethical utility comes in. In any antinomic structure, one can try to make the terms of the two theses appear in turn as the real entity reflected in the fictitious other, and consider what results from it.

Is it not the word *meeting*, as it is used now and as its use has developed over time (if there is any meaning in considering it as a set of *Schichten*, of strata), which indicates the true direction, the reality, the exact fold, in the opposed game of abstract concepts that become as many fictitious and unreal expressions? Is it not amazing that a part of the French word *rencontre*, and not the least important part, in Greek, is συμβολή, the famous symbol that allows two opposed parties to acknowledge that they can recreate the unity of the *tessera hospitalis* together? It might be said that contradictions are then brushed under the carpet, that one refuses to see that the words of the common language dangerously amalgamate them at the logical level, and that it would be better to criticize language in the name of concepts. Should not language, having been made neither for scientists nor philosophers, be severely questioned? In other words, is it possible to trust language, as it has become, at a certain stage of a given linguistic community, or should one rather prefer the logical game we have mentioned?

In fact, one does not have any choice, since the game of concepts, though it may become somewhat independent from language, still originates from a work of language on itself, or of the indefinite work of language by those who speak it. It is through a multitude of confrontations that it has become what it is. Though the Cartesian theory of whirlpools did not meet the success expected by the man who recommended it in physics, maybe it provides a better model for these innumerable and constant confrontations which result – provided that its unity is not too indeterminate and abstract – in the fact that in a language those original products which survive could hardly be different, and liable to correction, as it were from outside, by concepts. Once again, concepts themselves are the products of a particular treatment of these linguistic entities. One must therefore, start from language and its absolutely singular idiosyncrasies. It produces paths and traces which belong only to it, and which should be inherited in their transfer, admittedly not to keep them piously, which is not in any case possible, but in trying to understand and work on them in their singularity. *Intimacy* and *meeting* are among the paths cleared by numerous others, which one must at the same time follow, even as one stands alert to the need to correct their trajectory. One should not fear the multiplicity of paths in the different languages, for it is possible, from the bottom of a singular language, to seize the singularities of another, to distinguish the former from the latter, and to enrich oneself and grasp both in absolutely unheard-of ways. It is absolutely not a matter of concern that an ethics is not universal. It should not fear the singularities of either language or situations. Rather, they are each essential elements the one of the other. The one gives rhythm to the other. If morals may well claim formal universality, it is the very essence of ethics not to be able to do so, and to seek to solve only particular problems. It is therefore essential for ethical categories to be understood in movements that are inseparably existential and very singularly linguistic, and which cannot easily claim a general extension. Thus to the three or four traces or paths mentioned in the book of *Proverbs* in the Bible which lists them like Prévert, Borgès or Perrec did, and which have already been listed in the previous chapter,[21] one should add the path of ethical peregrinations.

Thus, to sum up, there are three fundamental reasons which permit one to prefer one solution to the other. The *first one* concerns the conception of languages in their relation to reason. When we give immediate and dogmatic preference to the concept, the different languages are assessed depending on the 'choices' they make in the antinomic structure, which is given as a sort of common denominator which should be imposed from the start. However, this approach is only acceptable if we posit a highly unlikely innate or transcendental reason for it. Reason is built up through a

[21]See p. 131–132, above.

more direct confrontation of words in each language, and of languages one against the other. Precisely however, when signs are preferred, the comparison of languages becomes more direct: we do not consider that truth must lie on an eternal conceptual *stele* which would make it possible to compare them. Rather, by starting from the complex configurations – which are foreign to each other – that languages are, we must shape the concepts which allow us to account for the differences.

We give primacy to the idiomatic singularity for *two further reasons* in addition to the previous conception of translation. *First*, languages are better springs of action than abstract concepts. Moreover, we better understand the motives of actions thanks to singular languages than thanks to abstract concepts. *Therefore*, it is more acceptable to treat languages as primary data than to consider abstract concepts themselves, by which one tries to understand their content. For concepts to be able to play that role of selection in languages, they should be secure, but they only become so upon the condition that some work on language ensures it. The work on language comes before the work from concepts.

Depending on the sinuous paths we have mentioned, ethics has no reason to be attuned with morals. The rules ethics requires could not be the same as those to which moral agents submit themselves, if only because moral rules concern only the matter of the ethical questions which need to be solved, without giving them their form. If, for example, the idea of *duty* may mean something in morals in which the good may appear to an agent who wants to act morally, it does make as much sense in ethics, except insofar as one *must* look for peace rather than conflict and war, but the latter condition is not even a general rule. First, because the good takes a plural form in ethics, and could not be one. Next, because, if there is one duty in ethics, it belongs more to the *necessity* of finding a solution to conflicts than to the specific imperative that Kant defined for moral duties. Detecting ethical categories may be done simply by borrowing from universalistic morals, as is the case in Kantianism, but only by that game of balances – that we have explained – of the theory of fictions which makes it possible to detect which propositions are interesting, and may pose reverse problems, as in mathematics. It is clear that the idea of *meeting* allows this game, as would that of *intimacy*, which we have analyzed in the previous volume,[22] and those of *risk* and *probability*, to which we have devoted several analyses in this one.

In ethics, psychoanalysis is an ally when solving reverse problems. It also takes the reverse path of folding through apagogical propositions, which is characteristic of the abstract critical approach, but also the philosophies that claim they are going beyond that enfolding by a synthesis which retains all the defects of abstraction of the apagogical enfolding.[23] Psychoanalysis throws itself onto the other end of that

[22]See *Rethinking Medical Ethics*, [4], pp. 139–160.

[23]Kierkegaard saw this point very well in his criticism of Hegel in *Repetition*: 'La reprise est proprement ce que l'on a, par erreur, appelé la *médiation*. Il est incroyable de voir quel bruit l'on a fait dans la philosophie hégélienne autour de la *médiation*, et le flot de niaiseries que l'on a mises en honneur sous cette enseigne. On eût mieux fait de soumettre à un examen rigoureux la notion de *médiation*. [. . .] Le terme de *médiation* est étranger; celui de *reprise* est un bon vieux mot danois, et

which the critical perspective deems not capable of being conceptualised, and proceeds to conceptualise it. However, that reversal into a concept of what the critical perspective does not pose in that form is a possibility that it gives itself, but it is not always acceptable. Psychoanalysis does not only consist in taking the view opposite to that of the critical perspective. Not every position that reverses the critical perspectives is a good one to take. One must separate, among singular terms borrowed from vernaculars, those which are worthy of that reversed conceptualization and those which are not. *Meeting* is undoubtedly one of those terms about which one should no longer hesitate.

One should not fear language difference in ethics. Admittedly, languages do not necessarily yield the same results, and a problem that is solved in one country, which has its own laws, morals and healthcare system, cannot necessarily be solved in the same way in another one, even a close one, even though the latter may draw inspiration from the solutions found in the former. We have shown, in a recent article, published by Bucharest University, that yielding on language in medical questions and medical ethics is not without consequences on the care of patients whose expression of pain in their idiomatic language is one of the conditions for an acceptable ethics. This was the intuition of Bentham's *Springs of Action*, a book published at the beginning of the nineteenth century which was little understood by his contemporaries [2] [3]: that to grasp the springs of action one needs to summon and employ languages in their most singular and untranslatable aspects. Languages point to the destinies of the springs of action and sensitivity. Bentham seized Hume's analysis of passions, or what the author of the book 'Of Passions' in *A Treatise of Human Nature*, called *impressions of reflection*, to show that the bouncing back that the idea of *reflection* represents compared to the idea of *impression*, was a fold that linked the structural reality of passions to economic, social, historical, political, ethical and 'moral' phenomena – to sum up the previous qualifications in one word – the meaning of which is different today. Bentham saw, in the Humean analysis of passions, the point where it was obvious that fictions could be substituted for them. It is difficult to appreciate the width or scope that Bentham gave to this insight. Maybe psychoanalysis, in particular in its Lacanian manifestation, ensures an essential and acceptable welcome to this sort of discourse today. There is a thought or understanding of language – of a language of which the words would not require to be corrected according to a meaning that preceded it in some concept that would transcend it, any more than language required one to be faithful to it, according to the unavoidable virtue of some hermeneutical conceptions. The aim is not to *listen* to the word, but to build its meaning, that is, to explain its authority over our thoughts and actions.

je fais compliment au danois de ce terme philosophique. De nos jours, on n'explique pas comment se produit la médiation; on ne dit pas si elle résulte du mouvement de deux moments antagonistes, en quel sens elle s'y trouve déjà contenue, ou si elle est un facteur nouveau qui intervient et, dans ce cas, comment' ([11], pp. 707–708).

Thus, when we insist on the fact that there is no empirical existence of the meeting, we do not of course mean that it does not have any individual or social effect, nor that its effect on human relations is insignificant. It is the exact contrary that is true. Words, in their idiomatic singularity, suck in behaviours, give them a direction, a structure, and capture them as in a well. They may even produce, up to a certain point, the effect of an ideal. One may be surprised to see that word becoming a humanist motto, to the point that a 'psychiatry of meeting' has been mooted.[24]

This is certainly not a proof that there is something like meetings, but it is evidence that one would like them to exist, which is already a certain mode of giving them a real effect. The meeting and the myth of the meeting are lived as a need, and they very efficiently produce the effects and the objects, or rather the images, they need.

If someone were surprised that the concept of *meeting*, which lacks any empirical corresponding element, while being absolutely singular, can nonetheless be linked to an ethical approach in which calculation forms an important dimension, we would refer back to what has already been said about intimacy, and which has already led us to considerations of the same type. We would add with two points. The *first* is that it is not because a concept cannot be empirically confirmed that it does not admit a rigorous mathematical approach. Pascal already noted this in relation to the idea of what he called, expressing his amazement in front of the title, 'la géométrie du hasard', which we could call *probability calculus*, in a famous dedication letter, *Celeberrimae Matheseos Academiae Parisiensi*, in which he boasted that the solution to the problem of the divisions, which, if it is 'maintenant [encore], [et] si elle a été rebelle à l'expérience, n'a pu échapper à l'empire de la raison. Car nous l'avons réduite en art avec une telle sûreté, grâce à la géométrie, qu'ayant reçu part à la certitude de celle-ci, elle progresse désormais avec audace, et que, par l'union ainsi réalisée entre les démonstrations des mathématiques et l'incertitude du hasard, et par la conciliation entre les contraires apparents, elle peut tirer son nom de part et d'autre et s'arroger à bon droit ce titre stupéfiant: *Géométrie du hasard*'.[25] The *second* is that calculation is not embarrassed by the singularity that arithmetic and geometry express, each in its own way, in their game with the universal; that a number, for example, allows us to differentiate with a maximum of precision two situations that are so close that geometry could hardly distinguish them, while geometry supposes a continuity between positions that numbers cannot but separate. There is a game of the universal and the singular in calculation, whether arithmetical or geometrical,

---

[24]It is often a pure effect of title, of a *captatio benevolentiae*, for, beyond the title, one discovers that no further mention of meeting appears, and that the so-called psychiatry is not very different from a classical one, at least it does not appear in a thematized form. As the word attracts a good press, it can bestow its blessing on the book without having to be analyzed. The Net provides, to whoever wants to consult it, a very large number of titles of books and symposia in psychiatry that include the word 'meeting' ('rencontre' in French) without its meaning being fully, or even adequately, explored in the body of the text. The term then functions as a kind of slogan for a humanist claim that has both virtues and limitations.

[25][16], vol. II, pp. 1031–1035.

which, far from hating singular affairs – such as those of ethics which one marks with the categories of intimacy, risk and meeting – is quite open and available to, and consonant with, them.

Bibliography

1. Bartabas. (2000). *Première Rencontre. Le Cheval et l'Homme: (Vingt Écrivains rêvent)*. Paris: Bartabas, Fernandez D., Desprez L., Phébus.
2. Bentham, J. (1983). *Deontology together with a table of the springs of action and article on utilitarianism* (A. Goldworth, Ed.). Oxford: Clarendon Press.
3. Bentham, J. (2008). *La Table des Ressorts de l'Action*. Paris: éd. de L'Unebévue.
4. Cléro, J.-P. (2019). *Rethinking medical ethics*. Stuttgart: Ibidem.
5. Descartes, R. (1982). The second meditation. In C. Adam & P. Tannery (Eds.), *Oeuvres de Descartes*. Paris: Vrin. IX-1.
6. Descartes, R. (1985). *The passions of the soul, the philosophical writings of descartes* (J. Cottingham, R. Stoothoff, & D. Murdoch, Trans., Vol. I). Cambridge/London/New York/New Rochelle/Melbourne/Sydney: Cambridge University Press.
7. Descartes, R. (1986). Les Passions de l'Âme. In C. Adam & P. Tannery (Eds.), *Oeuvres de Descartes*. Paris: Vrin. XI, IIde Part., 190 Art. LIII.
8. Descartes, R. (1996). Le Monde. In *Œuvres de Descartes* (Vol. XI). Paris: Vrin.
9. Ganascia, J.-G. (2017). *L'intelligence Artificielle*. Paris: Le cavalier bleu.
10. Hume, D. (2011). *A treatise of human nature* (D. F. Norton & M. J. Norton, Eds.). Oxford: Clarendon Press, 2 vols., Irst vol., Book 2.
11. Kierkegaard, S. (1993). *Ou bien ... ou bien, La reprise, Stades sur le chemin de la vie, La maladie à la mort*. Paris: R. Laffont.
12. Lacan, J. (1991). *Le Séminaire* (Livre VIII). Paris: Le transfert, éd. du Seuil.
13. Minkowski, E. (1933). *Le Temps Vécu*. Reproduced by PUF with the same pages in 2013.
14. Minkowski, E. (1936). *Vers une Cosmologie: Fragments Philosophiques*. Paris: Aubier.
15. Moore, G. E. (1912). *Ethics*. New York/London: Holt H./Williams & Norgate.
16. Pascal, B. (1970). Celeberrimae Matheseos Academiae Parisiensi, À l'illustre Académie parisienne de mathématiques. In *Oeuvres complètes* (Vol. II). Paris: Desclées de Brouwer.
17. Rousseau, J.-J. (1964). Du Contract Social ou Principes du droit politique. In *Oeuvres complètes* (Tome III, pp. 347–470). Paris: NRF Gallimard, Bib. de la Pléiade.
18. *Storytelling. Cabarets de Curiosités*. 2014. Dir. éd. Bardiot C. & Daurier R., Subjectile, Le Phénix scène nationale, Valenciennes.
19. Taulelle, D. (1984). *L'Enfant à la Rencontre du Langage*. Bruxelles: Mardaga.
20. Tisseron, S. (2015). *Les Robots Emphatiques*. Culture Mobiles, Octobre 2015.

Chapter 8
Epilogue

Abstract The epilogue added to the book was dictated by our sympathies with utilitarianism. A main prejudice lies in thinking that utilitarianism is a theory of happiness, or even of unbridled hedonism, inconsistent with the ethical issues we are dealing with today and which are seen as inherently global and longterm on earth and its surrounding space. We know that Bentham, at the end of his life, wanted to amend his utilitarianism in substituting systematically to the term « utility » -with which he had worked hitherto- the word « happiness » that would probably have spoiled the total body of his work. It seems that an invisible and salutary hand prevented him from making that modification; it was lucky. If I succeeded in clarifying it clear in the preceding chapters, one will understand that the main actor in ethics cannot be restrictively « the man »; that the necessity to settle the mixed being of technics and humanity -would it be overwhelmed by them- in largest ensembles of which the interest must prevail. Such a prevalence does not match necessarily with the satisfaction of humans -particularly when it is immediate and does not suffer to be postponed-. We think, with Lévi-Strauss, that man, even linked with digital and « intelligent » machines, will disappear before the end of the earth; and that it is difficult to avoid that the ethics of men coupled in such a way could do better than delaying their ending and accommodating this time for hopefully as long as possible and a maximum of generations with a pleasure of living that is not only *prima facie* and of first-degree.

Keywords Creation · Ecological wisdom · End of humanity · Lévi-Strauss C. · Nature · Promethean humanism · Religion

Our endeavour of having ethics gravitate around concepts that do not have any grand pretentions of determining the substance of its rights and duties by a straightforward process of deduction – which we have delineated in particular when dealing with *intimacy* and *meeting* – could of course be continued with the study of other concepts, which are often deemed to be minor ones, like *consolation*, *solidarity* and a few others. These ideas seem less robust and more fragile because they have not been studied as intensively as their relatives which have been taken care of in the

J.-P. Cléro, *Reflections on Medical Ethics*, Philosophy and Medicine 138,
https://doi.org/10.1007/978-3-030-65233-3_8

ethics that claims that it can deduce duties and rights. However, such frailty seems all the more important as it fits the inductive character of ethics, which is born from singular situations and from the necessity to resolve conflicts when they arise. We have mentioned psychoanalysis as a potentially useful resource for doing some research on these ideas. Resorting to psychoanalysis is understandable insofar as it is a practice involving two individuals which always happens in specific circumstances. When collecting these ideas, one might also consider the field of theologies. The problem, in their case, is that they have become too sensitive to the criticisms made by those who contest them or challenge their existence, so that, instead of generously offering singular ideas, they seem preoccupied with fending off the blows they receive. This is why the dictionaries of Christian ethics – and their examination does not imply that we necessarily agree to adopt such a concept,[1] as if one had to believe in God or at least assume God's existence, and even that of one God in particular, to think about ethics and practice it – seem to be afraid of mentioning consolation, for example. Either they do not mention it at all, or they use so many precautions in doing so that the idea becomes puny where one might have expected or hoped for a productive development.

One of the advantages of the ethics of virtues from this point of view, and despite all the defects it might have – namely that of shutting up what is said within predetermined limits, whereas ethics does not know any *a priori* – is due to the fact that it makes it possible to pour back into ethics that rich knowledge which has been carefully developed since Antiquity, and is threatened by the will to deduce everything from two or three abstract formulas.

This last point cannot be separated from a regret and an opening. We have talked about nature in terms of an essentially Promethean meaning, as if it did not exist without men's actions and we were its legislators. That is, at least, the reading we made of Kantianism, when we opposed it to Bayesianism, to suggest an ethics that would take more into consideration a probabilistic conception of action. We regret cleaving exclusively to that reading of nature, for there is another which, far from being an integration into today's human world characterized by culture and technical power over phenomena, is also – as one can become increasingly conscious today – that which folds in our actions, which gives them their horizon and their limit, rather than being that which constitutes their simple conditions of existence. This type of ethics is found in Lévi-Strauss especially, when he rightly claims he is a disciple of Rousseau. It is possible that such sorts of consideration may be crucial for ethics, in that almost none of our actions can happen without impacting on, and perhaps destroying, our environment. We are not advising that we should refrain from doing anything, but rather that we should be conscious, before and during any of our actions, of that potential destruction, so as to give them the best possible form, given their consequences for the earth.

Such an ethics, which adopts a low profile and features abstinence as a central value, of course meets at every turn the political or religious objections of a

[1] We have even explicitly rejected it.

Promethean humanism. In politics, whether one considers that the first value is the endeavour, whatever is being endeavoured, or instead the sharing of one's profits and benefits, does not alter the result much. In religion, one may make man dependent on God, but when one makes man the privileged partner of God the religious is not so different from an atheist humanism, except maybe in its vocabulary. The usual divisions of political debate – right or left – and those of the debate between believers and non-believers are not, from this point of view, the most important ones. In any case, human intelligence tends to place itself at the core of the complete system, and believes, or pretends it believes, that the whole universe can be organized around that belief. Lévi-Strauss's aim was to ponder the right that this puny agitation of our brains has to consider itself to be the centre of the things it imagines all exist to serve it. Exactly the opposite is true. Human intelligence, which is certainly a power of order and ordering, must be put in its less hegemonic place in an organization which is more important than it is, though not necessarily more powerful in the medium or long term. The orientation of intelligence – and especially human intelligence – must essentially be that of a supreme modesty, or even of self-effacement, since man may have to offer to dissolve himself.[2] At the least the engraving of his existence in the world should only be made by means that do not intensify entropy in the world into which it is inserted. Human intelligence must leave the other beings some space in order to understand itself as an other for those others, and not insist on taking up more space than would be entailed by that de-centring which, remote from our usage and habit though it may be, captures our true position and nature. It must abandon its ambition to impose upon the whole world the set of laws by which it is governed. Man is not the meaning of the world. The world does not have to align itself to the meaning that man intends to impose upon it according to almost all the dominant ideologies of our societies. The so-called destination of man does not have to conceive itself as the so-called destination of the world. Man may experience this quite vividly when he has aesthetic emotions.[3] One begins to perceive that Lévi-Strauss' nihilism or

[2]There is a sentence in *La pensée sauvage* (1962) which Foucault seized upon: 'Le but dernier des sciences humaines n'est pas de constituer l'homme, mais de le dissoudre' ([4], p. 326). At least the author of *Les mots et les choses* drew his inspiration from it in 1966.

[3]Man is not necessarily the most beautiful spectacle for himself. At the end of *Tristes Tropiques* ([8], p. 445), Lévi-Strauss talks about 'la contemplation d'un minéral plus beau que toutes nos oeuvres', '(la saisie d'un) parfum, plus savant que nos livres, respiré au creux d'un lis', '(celle d'un) clin d'oeil alourdi de patience, de sérénité, de pardon réciproque, qu'une entente (permet) involontairement parfois d'échanger avec un chat.' Like Kant, for whom beauty was a sort of sign of the moral value that he considered as being essentially interpersonal (*Critique du jugement*, § 42, § 59), LéviStrauss read natural beauty as the sign that ethics must not be centered upon man, and that even man knows it necessarily thanks to that sign. Lévi-Strauss's thought was decidedly the reverse of Kantianism. He did not hesitate to say that, thanks to ethnology, 'nous apprenons ainsi à mieux aimer et à respecter la nature et les êtres vivants qui la peuplent, en comprenant que les végétaux et animaux, si humbles soient-ils, ne fournissent pas seulement à l'homme sa subsistance, mais furent aussi, dès ses débuts, la source de ses émotions esthétiques les plus intenses et, dans l'ordre intellectuel et moral, de ses premières et déjà profondes spéculations' ([7], p. 166).

agnosticism had a positive side which becomes meaningful in a sort of ecological wisdom, according to which intelligence recognizes that it is folded and shaped by other orders which are as well or better founded than its own.

Lévi-Strauss' admiration for Rousseau has no other meaning than this,[4] and helps us to understand the mistake that is usually made in discussion of Rousseau's Epimethean values by Promethean humanism, as for instance by Fichte, who read them in a psychological manner – as concerned solely with the person of Rousseau – and denied any philosophical interest to what he interpreted as a domination of reason by affect,[5] although it is actually something quite different, which is precisely an intelligence which calculates a modest position which will allow it to persist in the least unstable position possible where it can find its happiness without the latter producing too much unhappiness or harm for other beings. Prometheanism blinded the moderns so much that it became invisible to them. They so strongly rejected the idea of *nature*, and so much favoured the movement of inserting nature into culture, that the simple project of inverting those values and acting so that culture, in turning away from its own imperial over-reach, is obliged to insert itself into nature, is no longer understood or is not felt as anything other than a violent contradiction. From this point of view, it is quite moving that Lévi-Strauss, who showed many signs of agnosticism in his book, and who dealt with religions in their symbolical dimension without seeming to believe in them in any way,[6] often used, though he weighed each of his words carefully, the word *création* (creation) to talk about the world or about nature as a set of beings.[7] His choice of this word is a warning, all the more so as it

[4]Derrida did not miss it when he wrote in *L'écriture et la différence* [1]: 'Si Lévi-Strauss, mieux qu'un autre, a fait apparaître le jeu de la répétition et la répétition du jeu, on n'en perçoit pas moins chez lui une sorte d'éthique de la présence, de nostalgie de l'origine, de l'innocence archaïque et naturelle, d'une pureté de la présence et de la présence à soi dans la parole; éthique, nostalgie et même remords qu'il présente souvent comme la motivation du projet ethnologique lorsqu'il se porte vers des sociétés archaïques, c'est-à-dire, à ses yeux, exemplaires. (...) Tournée vers la présence, perdue ou impossible, de l'origine absente, cette thématique structuraliste de l'immédiateté rompue est donc la face triste, *négative*, nostalgique, coupable, rousseauiste, de la pensée du jeu dont l'*affirmation* nietzschéenne, l'affirmation d'un monde de signes sans fautes, sans vérité, sans origine, offert à une interprétation active serait l'autre face.'

[5]Rousseau 'avait de l'énergie, mais plutôt l'énergie de la souffrance que celle de l'activité; il sentait fortement la misère des hommes; mais il sentait beaucoup moins la force propre qu'il avait pour porter aide à cette misère; et ainsi il jugea des autres de la même façon qu'il se sentait lui-même [...] Il tint compte de la souffrance; mais il ne tint pas compte de la force que l'humanité a en soi pour se secourir' (*La destination du savant, Cinquième Conférence*, [2], p. 103).

[6]*L'homme nu*, Final: 'À moi aussi sans doute, le domaine de la vie religieuse apparaît comme un prodigieux réservoir de représentations que la recherche objective est loin d'avoir épuisé; mais ce sont des représentations comme les autres, et l'esprit dans lequel j'aborde l'étude des faits religieux suppose qu'on leur refuse d'abord toute spécificité' ([6], p. 570).

[7]'Le droit à la vie et au libre développement des espèces vivantes encore représentées sur la terre peut seul être dit imprescriptible, pour la raison très simple que la disparition d'une espèce quelconque creuse un vide, irréparable à notre échelle, dans le système de la création.' *Tristes Tropiques* also mentioned creation: 'Les institutions, les moeurs et les coutumes, que j'aurais passé ma vie à inventorier et à comprendre, sont une efflorescence passagère d'une création par rapport à

does not refer to 'our creation', and underlines the necessity for us to adopt some strategy of submission, even though we do not have to find its modalities ourselves. We do not have to respect only reasonable beings, nor even only men, nor even living beings, but the world, which must be placed 'avant la vie, la vie avant l'homme', respect for the other living beings thus preceding respect for our own self-esteem. ([5], p. 422).

Bibliography

1. Derrida, J. (1967). *L'écriture et la différence*. Paris: Seuil.
2. Fichte, J. G. (2016). *La destination du savant*. Paris: Vrin.
3. Foucault, M. (2015). *Les Mots et les choses*, Paris, Gallimard, Bibliothèque de la Pléiade. *The order of things*. monoskop.org/images/a/a2/Ffoucault Michel The Order Things 1994.
4. Lévi-Strauss, C. (1962). *La Pensée sauvage*. Paris: Plon.
5. Lévi-Strauss, C. (1968). *L'origine des manières de table*. Paris: Plon.
6. Lévi-Strauss, C. (1971). *L'homme nu*. Paris: Plon.
7. Lévi-Strauss, C. (1983). *Le regard éloigné*. Paris: Plon.
8. Lévi-Strauss, C. (2008). *Tristes Tropiques*. NRF Gallimard.

laquelle elle ne possède aucun sens, sinon peut-être celui de permettre à l'humanité d'y jouer son rôlc' ([8], p. 443).

Bibliography

1. Derrida, J. (1967). *L'écriture et la différence*. Paris: Seuil.
2. Fichte, J. G. (2016). *La destination du savant*. Paris: Vrin.
3. Foucault, M. (2015). *Les Mots et les choses*, Paris, Gallimard, Bibliothèque de la Pléiade. *The order of things*, monoskop.org/images/a/a2/Ffoucault Michel The Order Things 1994.
4. Lévi-Strauss, C. (2008). *Tristes Tropiques*, NRF Gallimard.
5. Lévi-Strauss, C. (1962). *La Pensée sauvage*. Paris: Plon.
6. Lévi-Strauss, C. (1968). *L'origine des manières de table*. Paris: Plon.
7. Lévi-Strauss, C. (1971). *L'homme nu*. Paris: Plon.
8. Lévi-Strauss, C. (1983). *Le regard éloigné*. Paris: Plon.

Bibliography of the Quoted Authors

9. Aeschylus. (1988). *The Persians*, in *Aeschylus*, trans. H.W. Smyth, 2 vols., Cambridge, MA: Harvard University Press, London: W. Heinemann, vol. I.
10. Aristotle. (1991). *The poetics*. Cambridge/MA/London: Harvard University Press.
11. Aristotle. (1980). *The physics*, in 2 vol., I & II, Cambridge, MA: Harvard University Press, London: Heinemann W., The Loeb Classical Library, vol. IV–V.
12. Aristotle. (1989). *The metaphysics*, trans. H. Tredennick, in 2 vol., *Aristotle in twenty-three volumes*. Cambridge, MA: Harvard University Press, Cambridge University Press, London: The Loeb Classical Library, vol. XVII–XVIII.
13. Aristotle. *The poetics*. Cambridge, MA: Harvard University Press, London: The Loeb Classical Library, vol. XXIII.
14. Bachelard, G. (1991). *Le nouvel esprit scientifique*. Paris: Quadrige/PUF.
15. Bachelard, G. (1973). *L'Eau et les Rêves. Essai sur l'imagination de la matière*. Paris: Corti.
16. Bacon, F. (2006). *The essayes or counsels, civill and morall* (ed: Kiernan). Oxford: Clarendon Press. Particularly, Essay XVIII, Of Travaile.
17. Balibar, F. (1986). *Galilée, Newton lus par Einstein*. Paris: PUF.
18. Bardiot, C., & Daurier, R. (2014). *Storytelling. Cabarets de curiosités*, Subjectile, Le Phénix scène nationale, Valenciennes, .
19. Bartabas. (2000). *Première Rencontre. Le Cheval et l'Homme: (Vingt Écrivains rêvent)*, Partabas Fernandez D., Desprez L., Phébus, Paris.

J.-P. Cléro, *Reflections on Medical Ethics*, Philosophy and Medicine 138,
https://doi.org/10.1007/978-3-030-65233-3

20. Bayes, T. (2017). *Essai en vue de résoudre un problème de la doctrine des chances*. Paris: Hermann.
21. Bayes Th. *An Essay towards solving a Problem in the Doctrine of Chances*, in *Essai en vue de résoudre un problème de la doctrine des chances*, Cahiers d'Histoire et de Philosophie des Sciences, n° 18, -1988, Société Française d'Histoire des Sciences.
22. Bentham, J. (1843). *An Introduction to the Principles of morals and legislation*, in *The Works of Jeremy Bentham* (vol. I). Edinburgh: Tait W.
23. Bentham, J. (1970). *An Introduction to the Principles of morals and legislation* (eds: J. H. Burns & H. L. A. Har). London: Athlone Press.
24. Bentham, J. (1843). *Plan of parliamentary reform*, in *The works of Jeremy Bentham* (vol. III, ed: J. Bowring). Edinburgh: Tait W.
25. Bentham, J. (1843). *Pannomial fragments*, in *The works of Jeremy Bentham* (vol. III). Edinburgh: Tait W.
26. Bentham, J. (1843). *Constitutional code*, in *The works of Jeremy Bentham* (vol. IX, ed: J. Bowring). Edinburgh: Tait W.
27. Bentham, J. (1984). *Chrestomathia* (ed: M. J. Smith & W. H. Burston). Oxford: Clarendon Press.
28. Bentham, J. (1970). *Of laws in general* (ed: H. L. A. Hart). The Athlone Press.
29. Bentham, J. (2010). *Of the limits of the penal branch of jurisprudence*. Oxford: Clarendon Press.
30. Bentham, J. (2008). A table of the springs of action. In *Deontology together with A Table of the springs of action and article on utilitarism* (ed: A. Goldworth). Oxford: Clarendon Press, 1983. There exists a French version of this book: *La Table des Ressorts de l'Action*, Cahiers de l'Unebévue (éd: L'Unebévue). Paris.
31. Bentham, J. (1823). *Not Paul, but Jesus*, under the fictitious name of Gamaliel Smith, Esq., *Not Paul, but Jesus*. London: Hunt.
32. Bentham, J., *Rationale of Judicial Evidence* (5 vols.). London: Hunt & Clarke.
33. Bernoulli, J., *Ars conjectandi, pars quarta, traders sum & applications precedents doctrine in civilibus, moralibus & oeconomicis*, trad. de Meusnier N., IREM de Rouen, 1987, from the edition of Basel, published in 1713 to Thurnisien, reprint in *Die Werke von Jakob Bernoulli*, Birkhäuser, Basel, 1975, 3rd Vol., pp. 239–259.
34. *Bible*, Lévitique, in *L'ancien Testament*, vol. I, NRF, Bibl. de la Pléiade, 1970, Les Proverbes, in *L'ancien Testament*, vol. II, NRF, Bibl. de la Pléiade, 1967.
35. *Bible*, Actes des apôtres, épîtres aux Romains, épîtres aux Corinthiens, épître aux Éphésiens, épître aux Hébreux, épîtres de Jean, Apocalypse de Jean), in *Nouveau testament*. Paris: NRF Gallimard, 1971.
36. *The Holy Bible containing The Old and New Testaments*, Cambridge University Press, 1989.
37. *The Bible*, The Acts, The Epistle of Paul the Apostle to the Hebrews, The Epistle of Paul the Apostle to the Ephesians, The first Epistle General of Peter, The Gospel according Saint John; the Revelation of Saint John the Divine, in *The New Testament*, Authorized King James Version, The Gedeons International.
38. Bouveresse, J., « La théorie des fictions chez Bentham », in *Regards sur Bentham et l'utilitarisme*, éd. Mulligan K. et Roth R., Genève, Droz.
39. Boyle, R. (1996). *A free enquiry into the vulgarly received notion of nature* (eds: E. B. Davis & M. Hunter). Cambridge University Press.
40. Carr, N. (2019, February). « Un grand maître humilié redresse la tête », in *Books*, n° 94.
41. Cavaillès, J. (1987). *Sur la logique et la théorie de la science*. Paris: Vrin.
42. Cléro, J. P. (1985). *La philosophie des passions chez Hume*. Paris: Klincksieck.
43. Cléro, J. P. (1995). *Hume. Une philosophie des contradictions*. Paris: Vrin.
44. Cléro J.-P. (2001, avril). « Personne et anonymat. Du mauvais usage de la notion de personne », in *Les cahiers du Comité consultatif national d'éthique pour les sciences de la vie et de la santé*, n° 27 (pp. 35–38).
45. Cléro, J.-P. (2011). *Calcul moral ou Comment raisonner en éthique, A*. Paris: Colin.

46. Cléro, J.-P., « Réflexions critiques sur l'usage de la notion de personne en éthique médicale », in *Deux siècles d'utilitarisme*, sous la direction de M. Bozzo-Rey et d'É. Dardenne, Presses Universitaires de Rennes, 2e semestre 2011, pp. 211–231.
47. Cléro, J.-P. (2013). « La solidarité peut-elle se substituer à la valeur d'autonomie ? » Rouen: PURH.
48. Cléro, J.-P. (2015, janvier-mars). « Personne et altérité dans l'utilitarisme ». *Ethics, Medicine and Public Health*, 82–90, Sffem, Elsevier Masson. https://doi.org/10.1016/j.jemep.2014.09.001.
49. Cléro, J.-P. (2016, December). Has the care in psychiatry other characteristics than those it has in the other fields of medicine? *Revista Româneasca ventru Educatie Multidimentionala, 8*(2), 45–56. https://doi.org/10.18662/rrem/2016.0802.04.
50. Cléro, J.-P., (2016, décembre).« Y a-t-il, chez Stuart Mill, une spécificité de l'éthique entre les morales et le droit ? *Philosophical Enquiries, revue des philosophies anglophones*, n° 7.
51. Cléro, J.-P. (2016). 'Memento mori ou: Comment l'éthique, qui est une pensée de la vie, peut-elle inclure l'idée de la mort ? *Ethics, Medicine and public Health, 2*(2), 246–255.
52. Cléro, J.-P. (2017, juillet). Une pensée de l'existence à l'épreuve de l'éthique des soins. Les contradictions de l'éthique médicale. In revue *Cités*, Collectif *Jankélévitch; morale et politique*, *Cités*, n° 70. Paris: PUF.
53. Cléro, J.-P. (2018). « Qu'est-ce que soigner ? » ; « Le soin est-il une aide ? » ; « Qu'est-ce que l'autonomie ? » In *Le soin, l'aide, care, cure.* Rouen: PURH, en collaboration avec Annie Hourcade.
54. Cléro, J.-P. (2018). *Rethinking medical ethics.* Ibidem, Stuttgart.
55. Cléro, J.-P. (2018). Ethics and the increasingly english-speaking psychiatric tower of Babel. *Annals of the University of Bucharest*, Philosophy Series, LXVII, vol. 2.
56. Coumet, E. (1970, May–June). « La théorie du hasard est-elle née par hasard ? » *Annales Économies, Sociétés, Civilisation*, 25th year, 574–598.
57. Derrida, J. (1967). *L'écriture et la différence*. Paris: Le Seuil.
58. Desargues, G. (1951). *L'oeuvre mathématique de Desargues*. Paris: PUF.
59. Descartes, R. (1982). *Discours de la méthode*, in *Oeuvres de Descartes* (eds: C. Adam & P. Tannery). Paris: Vrin, VI.
60. Descartes, R. (2018). (?), *Discourse on Method* (1637), trails. by J. Veitch, (http://pinkmonkey.com/dl/library1/book0648.pdf), 2018 (?).
61. Descartes, R. (1982). *Méditations*, in *Oeuvres de Descartes* (eds : C. Adam & P. Tannery). Paris: Vrin, IX-1.
62. Descartes, R. (1985). *Objections and replies*, in *The Philosophical writings of Descartes* (vol. II, trans. J. Cottingham, R. Stoothoff, & D. Murdoch). Cambridge/London/New York/New Rochelle/Melbourne/Sidney: Cambridge University Press.
63. Descartes, R. (1989). *Principes*, in *Oeuvres de Descartes* (C. Adam & P. Tannery, IX-2). Paris: Vrin.
64. Descartes, R. (1986). *Le Monde*, in *Oeuvres de Descartes* (C. Adam & P. Tannery, XI). Paris: Vrin.
65. Descartes, R., *Règles pour la direction de l'esprit*, in *Oeuvres philosophiques*, T. I (1618-1637). Paris: Garnier, 2010. Translation in *The philosophical writings of Descartes* (vol. I, trans. J. Cottingham, R. Stoothoff, & D. Murdoch). Cambridge/London/New York/New Rochelle/Melbourne/Sidney: Cambridge University Press, 1985.
66. Descartes, R. (1986). *Les passions de l'âme*, in *Oeuvres de Descartes* (Adam & Tannery, XI). Paris: Vrin.
67. Descartes, R. (1985). *The passions of the soul*, in *The philosophical writings of Descartes* (vol. I, trans. J. Cottingham, R. Stoothoff, & D. Murdoch). Cambridge/London/New York/New Rochelle/Melbourne/Sidney: Cambridge University Press.
68. Diderot, D. (2000). *Lettre sur les aveugles à l'usage de ceux qui voient*. Paris: GF Flammarion.
69. Diderot, D. (1916). *The letter on the blind*, in *Diderot's early philosophical works* (trans: M. Jourdain). The Open Court Publishing Cy.

70. Diderot, D. (2004). *Éléments de physiologie*. Paris: Honoré Champion.
71. Diderot, D. (2011). *Fragments politiques échappés du portefeuille d'un philosophe*, textes établis et présentés par G. Goggi, Hermann, Paris.
72. Emanuel, E. J., & Patterson W. B. (1998, janv). *Journal of Clinical Oncology, 16*(1), 365–371.
73. Eschyle. (1964). *Les Perses*, in *Théâtre complet*. Paris: Garnier-Flammarion.
74. Fichte, J. G. (2016). *La destination du savant*. Paris: Vrin.
75. Foucault, M. (2015). *Les Mots et les Choses*, Gallimard, Bibliothèque de la Pléiade.
76. Foucault, M., *The order of things*, monoskop.org/images/a/a2/FfoucaultMichelTheOrder Things1994.pdf
77. Galileo Galilei (1967). *Dialogue concerning the two chief world systems* (trans: S. Drake). Berkeley/Los Angeles/London: University of California Press.
78. Galois, É. *Écrits et mémoires mathématiques* (éd: R. Bourgne et J.-P. Azra). Paris : Gauthier-Villars.
79. Ganascia, J.-G. (2017). *Intelligence artificielle: vers une domination programmée ?*(éd: Le Cavalier Bleu). Paris.
80. Gardies, J. L. (1987). *L'erreur de Hume*. Paris: PUF.
81. Goffi, J.-Y., Transhumanisme, academic version in Kristanek M., *L'Encyclopédie Philosophique*
82. Granger, G.-G. (1988). *La philosophie du style, O*. Paris: Jacob.
83. Harari, Y. N. (2014). *Homo Deus*. Cambridge: Cambridge University Press.
84. Harari, Y. N. (2017). *Homo deus. Une brève histoire de l'avenir* (A. Michel).
85. Hare, R. M. (1992). *Moral thinking, its level, method and point*. Oxford: Clarendon Press.
86. Hare, R. M. (2020). *Penser en Morale*. Paris: Hermann.
87. Hare, R. M. (1993). *Essays on Bioethics*. Oxford: Oxford University Press.
88. Harsanyi, J. (1980). *Essays on ethics, social behavior, and scientific explanation*. Dordrecht, Boston, Londres, Reidel.
89. Harsanyi, J. C., Morality and the theory of the rational behaviour. In H. Sen & B. Williams (Eds.), *Utilitarianism and beyond*. Cambridge/Paris: Cambridge University Press/Éditions de la maison des sciences de l'homme.
90. Hegel, F. W. (1939). *La Phénoménologie de l'Esprit*, trad. J. Hyppolite, Aubier, Paris.
91. Hegel, F.W. (2018). *The phenomenology of the spirit* (trans: T. Pinkard). Cambridge University Press.
92. Hilbert, D. (1971). *Les fondements de la géométrie*. Paris: Dunod.
93. Hilbert, D. (1930). *Grundlagen der Geometrie* (ed: Teubner). Leipzig.
94. Hirschmann, A. O. (1977). *The passions and the interests. Political arguments for capitalism before its Triumph*. Princeton: Princeton University Press.
95. Hottois, G. (1979). *L'inflation du langage dans la philosophie contemporaine: causes, formes et limites*, Éditions de l'Université Libre de Bruxelles, Faculté des Lettres, Brussels.
96. Hume, D. (2011). *A treatise on human nature* (2 vols, eds: D. F. Norton & M. J. Norton). Oxford: Clarendon Press.
97. Hume, D.(1991). *Traité de la nature humaine* de Hume, IId Livre sur *Les Passions*, GF-Flammarion, Paris, .
98. Hume, D. (1986). Of *suicide*. In P. Singer (Ed.), *Applied ethics*. Oxford/New York: Oxford University Press.
99. Hume, D. (2001). *De l'immortalité de l'âme*, in *Essais moraux, politiques et littéraires*. Paris: PUF.
100. Hume, D., *Of the immortality of the soul*, in Hume D., *Essays, moral, political, and literary* (2 vols, eds: T. H. Green & T. H. Grose). Scientia Verlag, Aalen, 1992, vol. II, PUF, Paris, 2001.
101. Hume, D. (1971). *L'Histoire Naturelle de la Religion, et autres essais sur la religion*. Paris: Vrin.
102. Hume, D. (1956). *The natural history of religion* (ed: H. E. Root). London.
103. Husserl, E. (1959). *Recherches Logiques*, T. I: *Prolégomènes à la logique pure*. Paris: PUF.

104. Jankélévitch, V. (1983). *L'irréversible et la nostalgie*. Paris: Flammarion.
105. Kant, E. (1997). *Critique de la Raison Pure*. Paris: Quadrige/PUF.
106. Kant, I. (2000). *Critique of pure reason*, in The Cambridge Edition of the Works of Immanuel Kant. Cambridge, New York, Melbourne: Cambridge University Press.
107. Kant, I. (2007). *Kant's Groundwork of the metaphysics of morals*, in *Practical philosophy*. Cambridge: Cambridge University Press.
108. Kant, I. (1996). *Groundwork of the metaphysics of morals*, in *Practical philosophy*, The Cambridge Edition of the Works of Immanuel Kant. Cambridge University Press.
109. Kant, I. (1996). *The metaphysics of morals*, in *Practical philosophy*, trans. M. J. Gregor, Intr. by A. Wood) The Cambridge Edition of the Works of Immanuel Kant, Cambridge University Press, pp. 353-603.
110. Kierkegaard, S. (1993). *Ou bien ... ou bien, La reprise, Stades sur le chemin de la vie, La maladie à la mort*. Paris: R. Laffont.
111. Kierkegaard, S. (1967). *Les Miettes philosophiques*, trad. de P. Petit, Éditions du Seuil, Paris.
112. Kierkegaard, S. (1948). *Riens philosophiques*, trad. de K. Ferlov et de J. Gateau, Gallimard NRF, Paris.
113. Kierkegaard, S. (1985). *Philosophical fragments*, in Kierkegaard's Writings, VII, ed. & trans. by H.V. Hong & E. Hong. Princeton: Princeton University Press.
114. Lacan, J. (1978). *Le Séminaire*, L. II, *Le moi dans la théorie de Freud et dans la technique de la psychanalyse*. Paris: Le Seuil.
115. Lacan, J. (1986). *Le Séminaire*, L. VII, *L'Éthique de la psychanalyse The ethics of psychoanalysis*. Paris: Le Seuil.
116. Lacan, J.(1991). *Le Séminaire*, L. VIII, *Le transfert*. Le Seuil.
117. Lacan, J. (2004). *Le Séminaire*, L. X, *L'angoisse*. Paris: Le Seuil.
118. Lacan, J. (2001). « Maurice Merleau-Ponty », *Autres écrits*, éd. du Seuil, Paris, pp. 175–184.
119. Leibniz, G. W. (1978). *Méditations sur la connaissance, la vérité et les idées*, in *Opuscules choisis*, trad. by P. Schrecker. Paris: Vrin.
120. Leibniz, G. W. (1978). *Remarques sur la partie générale des Principes de Descartes*, in *Opuscules choisis*. Paris: Vrin.
121. Leibniz, G. W. (1989). *Critical thoughts on the general part of the principles of descartes* (1692), link.springer.com>content>pdf. In *Philosophical Papers and Letters*, ed. L.E. Loemker, Kluwer Academic Publihers, Chap. 42, 1976.
122. Leibniz, G. W. Animadversiones on descartes principles of philosophy, Books 1 and 2, *The philosophical works of Leibnitz, comprising The monadology, new system of nature, principles of nature and grace, letters to Clarke, Refutation of Spinoza and his other important opuscules, together with the Abridgment of the Theodicy, and extracts from the new essays on human understanding*, Tutthe, Morehouse & Taylor. (http://www.archive.org/steam/philosophicalwor00leibuoft_djvu.text).
123. Lévi-Strauss, C. (2008). *Tristes Tropiques*, NRF Gallimard.
124. Lévi-Strauss, C. (1962). *La Pensée sauvage*. Paris: Plon.
125. Lévi-Strauss, C. (1968). *L'origine des manières de table*. Paris: Plon.
126. Lévi-Strauss, C. (1971). *L'homme nu*. Paris: Plon.
127. Lévi-Strauss, C. (1983). *Le regard éloigné*. Paris: Plon.
128. Locke, J. (2010). *Some thoughts concerning education*. Cambridge/London: Cambridge University Press.
129. Locke, J. (1992). *Quelques pensées sur l'éducation*. Paris: Vrin.
130. Marx, K. (1993). *Le Capital, Critique de l'économie politique* (4ème éd.). L. Ier, Paris:, PUF.
131. Mauss, M. (2012). *Essai sur le don, Forme et raison de l'échange dans les sociétés archaïques*. Paris: Quadrige PUF.
132. Mauss, M. (2016). *The gift*, selected, annoyed, and translated by J. Guyer. Chicago: Hau Books.
133. Medical robotics, edited by J. Troccaz, Lavoisier, Cachan, 2012.
134. Merleau-Ponty, M. (1967). *La structure du comportement*. Paris: PUF.

135. Merleau-Ponty, M. (1945). *La Phénoménologie de la Perception*. Paris: NRF-Gallimard.
136. Stuart, M. J. (1972). *Collected works of John Stuart Mill*, vol. XVII, *The later letters of John Stuart Mill* (1849-1873). London/Toronto: Routledge & Kegan P.
137. Stuart, M. J. (1996). *On liberty*, in *Essays on politics and society, collected works of John Stuart Mill*, vol. XVIII. London/Toronto:Routledge & Kegan P., 1977.
138. Mill, J. (2014). *Stuart, Considérations sur le gouvernement représentatif*. Paris: Hermann.
139. Minkowski, E. (1936). *Vers une cosmologie: fragments philosophiques*. Paris: Aubier.
140. Minkowski, E. (2013). *Le temps vécu*. Paris: Quadrige PUF.
141. Montaigne M. de (1967). *Oeuvres complètes*, NRF Bibl. de la Pléiade.
142. Montaigne M. de (1927). *The essays of Montaigne* (trans: J. M. Robertson, 2 vols.). London: Oxford University Press/Humphrey Milford.
143. Montesquieu (1995). *De l'Esprit des lois* (vol. I & II). Paris: Gallimard.
144. Moore, G. E., *Ethics*. New York: Holt H. London: Williams & Norgate, 1912. Oxford: Clarendon Press, 2005.
145. Moore, G. E. (2019). *Éthique*. Paris: Hermann.
146. Moore, G. E. (1993). *Principia Ethica*, revised edition by Th. Baldwin, Cambridge University Press.
147. Neumann, I. B. (2015). *Diplomacy and the making of world politics* (ed: O. J. Sending). Cambridge University Press.
148. Nietzsche, F. (1995). *Humain, trop humain*. Paris: Le Livre de Poche.
149. Nietzsche, F. (1975). *Humain trop humain*. Denoël-Gonthier, Loos-lez-Lille, 2 vol.
150. Nietzsche, F. (1910). *Human, all-too-Human*, §§ 88, 8 (trans: H. Zimmern & T. N.Foulis). Edinburgh, London.
151. Nietzsche, F. (1992). *Ainsi parlait Zarathoustra*. Paris: Gallimard.
152. Nietzsche, F. (2013). *Thus spoke Zarathustra. A book for all and none* (eds: A. del Caro & R.B. Pippin, trans: A del Caro). Cambridge University Press.
153. Nietzsche, F. (1985). *Le Crépuscule des Idoles, suivi de: Le cas Wagner*. Paris: GF Flammarion.
154. Nietzsche, F. *Twilight of the idols*. http://www.inp.uw.edu.pl/mdsie/Political_Thought/twilight-of-the-idols-friedriech-nietzsche.pdf.
155. Nietzsche, F., *La Volonté de Puissance*, T. 2, transe. by H. Albert, § 256. books.Google.fr/books?id=HOswDAAAQBAJ&pg=PT21332Ipg=PT2133&dq=Nietzsche.
156. Pascal, B. (1992). *Oeuvres complètes*, I-IV, Desclée de Brouwer, Paris, (I) 1964, (II) 1970, (III) 1991, (IV).
157. Pascal, B. *Celeberrimae Matheseos Academiae Parisiensi* (1654), in *Oeuvres Complètes*, Bibliothèque européenne Desclée de Brouwer, Paris, 1970, vol. II.
158. Pascal, B. (2004). *Provinciales, Pensées*, La Pochothèque, Le Livre de Poche/Classiques Garnier, Paris.
159. Pascal, B. (1995). *Pensées*, trad. A.J. Krailsheimer. London: Penguin Books.
160. Passamani, E., *The New England Journal of Medecine* (vol. 324, N° 22, May 30th 1991, pp. 1585–1589).
161. Philonenko, A. (1997). *L'oeuvre de Kant – La Philosophie critique*, vol. II, *Morale et Politique*(3rd ed). Paris: Vrin.
162. Pilet, F., & Lelièvre, F. (2013). *Krach Machine*. Paris: Calmann-Lévy.
163. Plato. (1990). *Phaedrus*, in *Plato, Euthyphro.Apology.Crito.Phaedo.Phaedrus*, transl. by H.N. Fowler, The Loeb Classical Library, vol. I, Cambridge, MA: Harvard University Press, London: Heinemann W.
164. Plato. (1991). *Gorgias*, in *Plato, Lysis.Symposium.Gorgias*, The Loeb Classical Library, vol. I. Cambridge, MA/London: Harvard University Press.
165. Plato. (2013). *Republic* (Books 1-5). Cambridge, MA/London: Harvard University Press.
166. Plato (2013). *Republic* (Books 6-10) (ed & trans: C. Ellyn-Jones & W. Preddy). Cambridge, MA/London: Harvard University Press.
167. Plato. (1984). *The laws I & II*, in *Plato in twelve volumes*, X & XI. Cambridge, MA/London: Harvard University Press/Heinemann W.

168. Plato. (1986). *Epinomis*, in *Plato in twelve volumes*, XII. Cambridge, MA/London:Harvard University Press/Heinemann W.
169. Poe, E. A. (1951). *Aventures de Goldon Pym*, in: *Oeuvres en prose*, traduits par Baudelaire, NRF La Pléiade.
170. Poe, E. A. (1984). *Poetry and tales*. New York: The Library of America, Literary Classics of the United States.
171. Pouliot, V. (2015). *Diplomacy and the making of world politics* (ed: O. J. Sending). Cambridge University Press.
172. Rawls, J. (1999). *A theory of justice*. Oxford: Oxford University Press.
173. Ricoeur, P. (1988). *La philosophie de la volonté*. Paris: Aubier (Vol. I: Le volontaire et l'involontaire; vol. II: Finitude et culpabilité).
174. Rousseau, J.-J. (1964). *Oeuvres complètes*, vol. III, *Du Contract Social ou Principes du droit politique, Écrits politiques*, Gallimard, NRF, Bibliothèque de la Pléiade.
175. Rousseau, J.-J., *The social contract* (trans: J. Bennett).
176. Rousseau, J.-J. (1969). *Émile ou de l'éducation*, in *Oeuvres complètes*, IV. Paris: NRF Gallimard.
177. Rousseau, J.-J., *Emile or education* (trans: B. Foxley) London/Toronto/New York: J. M. Dit/E. P. Dutton.
178. Sartre, J.-P. (1943). *L'être et le néant*. Paris: NRF Gallimard.
179. Sartre, J.-P. (1984). *Being and nothingness* (trans: H. E. Barnes). London: Routledge.
180. Schiller, F., & Werke, L. C. Hanser, in 5 vols, Darmstadt, vol. I, 1987.
181. Sending, O. J., Pouliot, V., & Neumann, I. B. (2015). *Diplomacy and the making of world politics* (ed: O. J. Sending). Cambridge University Press.
182. Sextus Empiricus, in four volumes, III, *Against the ethicists* (trans. Rev. R.G. Bury). Harvard University Press/W. Heinemann, Cambridge, MA/London, The Loeb Classical Library, 1987, pp. 383–509.
183. Shapin, S., '« Homo deus » vraiment ?' In *Books*, fév. 2019, n° 94, pp. 30–35. S. Shapin's article, translated by D. Veaudor had already been published in the *London Review of Books* on July 13th 2017.
184. Sicard, D., To President Hollande on 18th December 2012, and was entitled *Penser solidairement la fin de vie <Thinking about the End of Life in Solidarity>*, 2012.
185. Sicard, D. (1999). *Hippocrate et le scanner*. Paris: Desclée de Brouwer.
186. Simondon, G. (1958). *Du mode d'existence des objets techniques*. Paris: Aubier.
187. Singer, P. (1986). *Applied ethics*. Oxford/New York: University Press.
188. Singer, P. (1993). *Practical ethics* (2nd ed.). Cambridge/New York/Melbourne: Cambridge University Press.
189. Spinoza, B. (1999, janv.). *Éthique*, éd. du Seuil, Paris.
190. Spinoza, B., *Ethics* (*Ethica ordine geometrico demonstrata*), trans. from the Latin by R.H.M. Elwes, globalgreyebooks.com.
191. *Storytelling. Cabarets de Curiosités*, dir. éd. BardiotC. & Daurier R., Subjectile, Le Phénix sème nationale, Valenciennes, 2014.
192. Sureau, C. (2006). L'inconnu dans la maison. In *Cités* n° 28, Paris: PUF.
193. Symposium of Paris-Descartes University on *La Vente des Produits Sanguins, Précarité et Vulnérabilité*, Feb. 8, 2018.
194. Taulelle, D. (1984). *L'enfant à la rencontre du langage*. Bruxelles: Mardaga.
195. Thomson, J. J. A defense of Abortion. In Singer P. (Ed.), *Applied ethics* (pp. 37–56).
196. Tisseron, S. (2015, October). *Les robots emphatiques*, Culture Mobiles.
197. Tooley, M. Abortion and infanticide. In P. Singer (Ed.), *Applied ethics* (pp. 57–85).
198. Troccaz, J. (2012). *Robotique médicale*, sous la direction de J. Troccaz, Cachan, Lavoisier.
199. Villani, C. (2012). *Théorème vivant*. Paris: Grasset.
200. Vuillemin, J. (1984). *Nécessité ou contingence. L'aporie de Diodore et les systèmes philosophiques*. Paris : Les Éditions de Minuit.
201. Zelizer, V. (2018). *The social meaning of market*. Princeton: Princeton University.

Name Index

J.-P. Cléro, *Reflections on Medical Ethics*, Philosophy and Medicine 138,
https://doi.org/10.1007/978-3-030-65233-3

Subject Index

J.-P. Cléro, *Reflections on Medical Ethics*, Philosophy and Medicine 138,
https://doi.org/10.1007/978-3-030-65233-3

U

V

W

The manufacturer's authorised representative in the EU is Springer Nature Customer Service Centre GmbH, Europaplatz 3, 69115 Heidelberg, Germany. If you have any concerns regarding our products, please contact ProductSafety@springernature.com

Printed and bound by CPI Group (UK) Ltd, Croydon, CR0 4YY
29/04/2026
02099460-0004